Automotive Brake Disc Materials

**Costel Dorel FLOREA, Costica BEJINARIU,
Nicanor CIMPOESU, Ramona CIMPOESU**

Gheorghe Asachi Technical University of Iasi, Romania, Materials Science and
Engineering Faculty

Published by **Materials Research Forum LLC**
Millersville, PA 17551, USA

Published as part of the book series
Materials Research Foundations
Volume 105 (2021)
ISSN 2471-8890 (Print)
ISSN 2471-8904 (Online)

Print ISBN 978-1-64490-144-1
ePDF ISBN 978-1-64490-145-8

This book contains information obtained from authentic and highly regarded sources. Reasonable efforts have been made to publish reliable data and information, but the authors and publisher cannot assume responsibility for the validity of all materials or the consequences of their use. The authors and publishers have attempted to trace the copyright holders of all material reproduced in this publication and apologize to copyright holders if permission to publish in this form has not been obtained. If any copyright material has not been acknowledged, please write and let us know so we may rectify in any future reprint.

Distributed worldwide by

Materials Research Forum LLC
105 Springdale Lane
Millersville, PA 17551
USA
http://www.mrforum.com

Printed in the United States of America
10 9 8 7 6 5 4 3 2 1

Table of Contents

Automotive Brake Disc Materials   Materials Research Forum LLC
Materials Research Foundations **105** (2021)   https://doi.org/10.21741/9781644901458

Chapter 1

Current Status of Materials used in the Construction of Vehicle Brake Discs

C.D. Florea, C. Bejinariu, N. Cimpoesu, R. Cimpoesu*

Faculty of Materials Science and Engineering, "Gheorghe Asachi" Technical University of Iasi, Romania

ramona.cimpoesu@tuiasi.ro

Abstract

Regardless of the method of propulsion, use or size, all vehicles have braking systems. Most braking systems currently use metal discs to perform the vehicle deceleration stage. The current requirements involve maintaining properties at high temperatures (high coefficient of friction, corrosion resistance, good thermal conductivity, etc.), when applying a load on the braking torque in motion with a certain sliding speed. This chapter presents the main knowledge about the materials used in braking systems and the possibilities to improve their properties.

Keywords

Braking Systems, Friction Materials, Tribology

1.1 Car- braking systems

Over the years, the field of materials for braking systems has evolved greatly due to the growing need to meet more solicitant operating requirements. The working conditions are constantly demanding with an increase in the weight of the bodies involved in the braking process, as well as in their movement speeds. In this sense, operating temperatures have escalated from 200 °C in the 1920s to 1000 ÷ 1200 °C in recent decades (in the case of jet aircrafts) [1].

Iron-based and copper-based friction materials are two classes of materials used in the manufacture of components in friction couplings. The book proposes the obtaining, through metallurgy-specific processes, of friction materials with high tribological characteristics, in the conditions of low energy consumption and considerably lower material losses. The materials used in the manufacture of braking systems form a friction coupling which should ensure the conversion of the system's kinetic energy into thermal energy via the friction surfaces.

Brake discs are parts used to slow down or stop a wheel from rotating. These brake disks are made of cast iron-carbon alloys, but in many cases can be made of composite materials (reinforced carbon-carbon, ceramic compounds, etc.).

A set of brake disks is connected to the wheel or axle. To stop the wheel, the friction material (brake pads mounted in a caliper) are forced mechanically, hydraulically, pneumatically or electromagnetically, to adhere to the surface of the brake discs. Friction causes the brake discs attached to the wheel to slow down or stop. The brakes convert the movement into heat, and when they become too hot, their efficiency decreases, appearing the phenomenon called *brake fade*. To slow down or stop a vehicle the kinetics and any potential energy of the vehicle should be taken into account. Recently, fuel efficiency and at the same time the attempt to reduce associated gas emissions has become one of the main goals of the automotive industry, and to this end some manufacturers started to produce electric and hybrid vehicles already available on the market.

These vehicles use an electric motor either as the main propulsion source or as a secondary source to aid the traditional internal combustion engine. A significant fuel economy can be achieved by converting some of the energy lost during braking, into electricity. This energy can then be stored in batteries and used for the propulsion of the vehicle, or for various power-consuming accessories such as air conditioning, lights, etc. However, it is not yet possible to recover more than $10 \div 15\%$ of the total energy used in propulsion [2], and therefore these vehicles also contain traditional braking systems as a safety reserve. The braking systems act to stop the vehicle by transforming the energy of the moving vehicle into heat and dissipating it into the surrounding atmosphere, and as a result, energy is lost without any chance of recovery. Despite this, the current use of a relatively small number of electric and hybrid vehicles means that brakes with the classic friction system are the dominant method of car stopping and will continue to be in the near future. Therefore, researches continue on ways and means to improve this technology in the field, such as weight loss, increasing heat dissipation, increasing wear resistance and coefficient of friction and improving their safety systems.

Friction brakes operate by transforming the kinetic and the potential energy of the vehicles into thermal energy (heat). This heat is created as a result of friction at the interface between a rotor (disc or drum) and stator (pads or brake pads). During braking, a large amount of heat can be created and the rotor will absorb it. The rotor and the surrounding components effectively operate the temporary thermal storage devices, and a sufficient cooling of these components is essential to achieve satisfactory performance of the braking system, [3].

Therefore, it is essential to dissipate the heat efficiently for the successful operation of a braking system.

There are two main types of car brakes, drum and disk brakes. Drum brakes operate by pressing obstacles radially outward on a rotating drum (rotor), while the brake disks operate by compressing axially the buffers (stators) against a rotating disc (rotor), Fig.1.1. A more advanced form of disk brake, used in practice recently, is the ventilation of this system to increase the dissipation capacity or the disc with holes, when the air causes internal cooling, entering through the radial passages or blades on the disk.

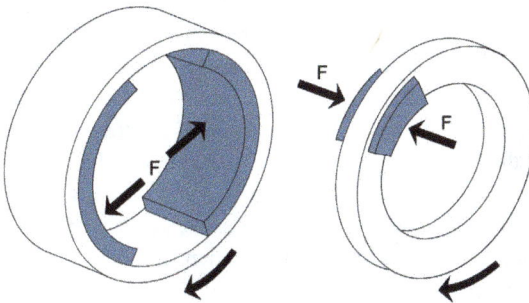

Figure 1.1. Scheme drum and disc brakes

Disc brake systems have different advantages compared to drum-based ones observed by universal adoption in cars as well as in front of light trucks [4, 5]. However, drum brakes are still used in many applications, including heavy trucks. The main advantages of the disk compared to the use of the drum are:

- the friction surfaces of the brake disc are exposed to an atmosphere that provides better cooling and reduces the possibility of thermal failure;
- in the case of drum brakes, the extension of the drum at high temperatures will result in a longer pedal stroke and improper contact between the drum and obstacles, while in disc brakes high temperatures increase in disc thickness, without any direct adverse effect in the braking system;
- the adjustment of the disc brake is done automatically, while the drum brakes will need to be adjusted for the used friction material;
- disc brakes are less sensitive to high temperatures and can operate in high safety conditions at temperatures up to 1000 °C. It is recommended that drum brakes, due to their geometry and the effects of the coefficient of friction, do not exceed the temperature of 500-600 °C [5].

Brake discs (both solid and hollow) are designed from an iron alloy and mechanically machined to make the final shape. Reducing fuel consumption and greenhouse gas emissions are a current and most serious issue in the automotive industry. The use of aluminium has significantly increased in the automotive recently, to reduce the weight of cars and improve fuel efficiency. As the brake disc or rotor is an essential component in terms of vehicle safety, the materials used for braking systems must have excellent friction and wear properties. They should be stable and reliable under different load conditions, speed, temperature and environment, but also have a high durability. Several factors may be taken into consideration selecting the material for the brake discs.

The most important aspect is the ability of the brake disc material to withstand high friction and less abrasive wear. Another requirement is to withstand high temperature that occurs due to friction. Weight, capacity of the manufacturing process and cost are also important factors in the design phase. In the material selection stage, the recyclability of cast iron is advantageous, but the evolution of CO_2 emissions during remelting must also be taken into account. The brake disc must have sufficient thermal storage capacity to prevent distortion or cracking of the material due to thermal stresses, for a time, until the heat can be dissipated. This situation is not particularly important for a single stop but is essential in the case of repeated stops at high speed.

Disc brake systems generate the braking force by attaching the brake pads to a rotor mounted on the hub. The major mechanical advantage of hydraulic or mechanical disc brakes is that it allows a small lever actuation force to be transformed into a high clamping force on the wheel. This force tightens the rotor with the brake pads and generates a high braking power [6]. The higher the coefficient of friction for the brake pads is the higher the braking power will be generated. The coefficient of friction may vary depending on the type of material used for the brake disc. Typically, brakes refer to the dynamic coefficient of friction or the coefficient of friction measured while the vehicle is moving [6].

All modern disc brake systems rely on brake pads pressed on both sides of the brake disc to increase rolling resistance and slow down the vehicle. The braking system is a vital safety component of ground transport systems; therefore, the structural materials used in brakes should have a combination of properties such as good compressive strength, higher coefficient of friction, wear resistance, low weight, good thermal capacity and economic viability [7, 8].

The most commonly used materials for braking systems are cast irons. In addition, in most cases at the laboratory level, titanium-based materials, aluminum matrix composites and ceramic inserts or only ceramic materials were tested. The cheapest option, with industrial applications in more than 95% of cases is also based on cast iron [9, 10].

1.2 Analysis of car-braking systems

The braking system is the most important part of all systems that make up a modern vehicle. This system is responsible for stopping a heavy vehicle moving at a high speed over a relatively short distance. The lives of traffic participants depend on the accuracy of this operation. Thus, we must ensure that is works correctly in terms of the properties of the metallic materials involved. This will not be possible without a correct understanding of the braking operation.

1.2.1 Tribological elements

Tribology is the science that deals with the interaction between two contact surfaces in relative motion. The term comes from the Greek „tribos", which means friction and „logos", meaning knowledge [11]. There are studied problems related to friction, wear and lubrication of the mechanisms. Friction and wear occur in most applications starting from the natural ones (example: wear of human teeth) and ending with those of high technologies (applications for space shuttles) [12]. Tribological research reduces material wear and implicitly increase the service life of mechanical and mechatronic systems, as well as friction control. These two objectives involve combining surface mechanics with physico-chemical knowledge, related to surfaces and interfaces in contact.

Friction is one of the most studied phenomena due to its presence in all moving mechanical assemblies [13]. The term comes from the Latin friction, which means "to rub" [11]. Friction is a complex process that occurs at the interaction of two moving bodies relative to each other, as opposed to movement [14]. According to the literature, friction can be classified into four groups:

a)   dry friction: the two surfaces of the coupling are in direct contact in the absence of any contamination of the contact surfaces or in the presence of a gaseous environment and reduced contamination of the contact surfaces; in this case the values of the coefficient of friction are high;

b)   limit friction: between the two surfaces of the dome there are thin, molecular layers formed by establishing in the surface layer either Van der Waals type bonds (adsorption) or chemical bonds, with electron exchange (chemisorption); the layer formed is continuous, preventing direct contact between the elements of the coupling;

c)   fluid friction: in this case, the two surfaces of the coupling are separated by a continuous film of lubricant;

d)   mixed friction: occurs at the limit of fluid friction, the lubricant film being interrupted from place to place by the peaks of the roughness of the surfaces in contact.

The device used to perform the determinations of the coefficient of friction called tribometer was continuously improved (Fig. 1.2). There are four laws that apply to dry rubbing. According to them, the friction force:

1) is directly proportional to the normal load;
2) does not depend on the size of the surfaces in contact;
3) does not depend on the relative speed of the surfaces in contact;
4) depends on the materials in contact [15].

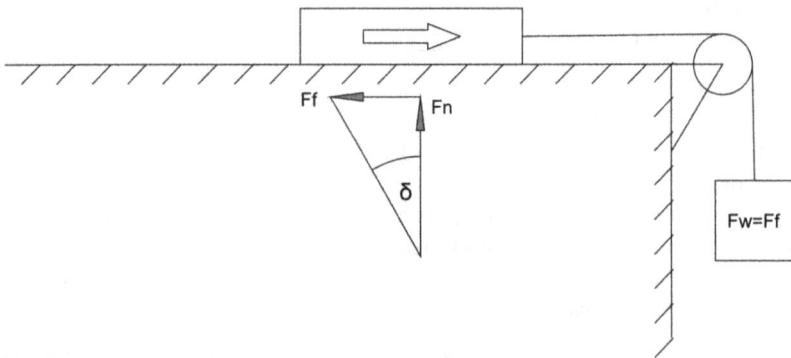

Figure 1.2. The tribometer, schematic, used for friction measurements

The Bowden-Tabor law applies for two rough metal surfaces and states that the real contact area that occurs as a result of the action of a normal force, F_n, depends on the ratio F_n/p_c., where p_c represents the flow pressure of the softer material. Friction is characterized by two parameters:

1)   friction force, F_f: directly proportional to the normal reaction, F_n, and independent of the contact surface area;
2)   the friction coefficient, μ: dimensionless quantity, which depends on the friction regime, is determined by the ratio between the friction force, F_f, and the normal reaction, F_n.

Wear, the effect of the friction, is a complex process in which a part of the material is removed and the initial state of the surfaces in contact changes. We have several researches on this process, that aimed at establishing the mechanisms through which it occurs, as well as identifying solutions to reduce it [16]. A material from a solid surface can be removed

Materials Research Forum LLC
https://doi.org/10.21741/9781644901458

by three methods: melting, chemical dissolution and physical separation of some atoms from the surface.

The classification of this process divides the wear into:

a)   adhesion (adherence) wear: this occurs by welding and breaking the welding bridges between the contact micro-areas (Fig. 1.3 left); in this case both the coefficient of friction and the wear intensity are high [17];

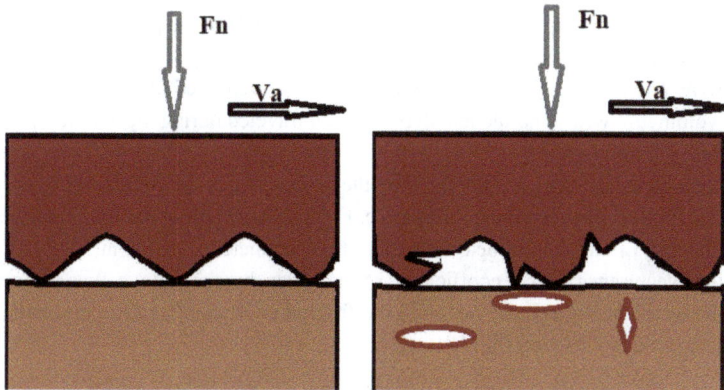

Figure 1.3. Adhesion wear scheme

b)   abrasion wear: this type of wear occurs due to the presence of hard particles between the surfaces in contact or due to the harder roughness of one of the surfaces;

c)   fatigue wear: in this case, the contact surfaces are subjected to cyclic stresses that cause plastic deformations, cracks, exfoliations or pinches in the surface layer (Fig.1.3 right);

d)   impact wear: occurs in equipment, installations or machine parts that in their operation are accompanied by impact phenomena [17];

e)   corrosion wear: appears due to the action of aggressive chemical factors from the environment in which the couple operates and due to the mechanical stresses to which the assembly is subjected [15].

Wear can be characterized by the following parameters:

1)   linear wear, U_h: is determined by measuring the length perpendicular to the surface to be worn, h;

2) mass wear, U_m: is determined by the difference between the initial mass of the sample and the mass of the sample after performing the wear test;

3) volumetric wear, U_v: the volume of material removed is determined;

4) planimetric wear, U_p: is determined by the calculation of area of the normal section applied on the wear surface;

5) wear resistance, R_u: represents the inverse of linear, mass, volumetric or planimetric wear;

6) wear speed: represents the ratio between the removed material (linear, mass, volumetric or planimetric wear) and the time in which the process occurs; the wear speed can be linear, mass, volumetric or planimetric;

7) wear intensity: represents the ratio between the removed material (linear, mass or volumetric wear) and the length of the road traveled during the test; it can be linear, mass or volumetric;

8) the wear coefficient, k: is defined by the ratio between the volumetric wear, U_v, and the product between the normal force, F_n, and the length of the road traveled during the test, L_f; the unit of measurement of this parameter is $mm^3/N \cdot m$;

9) the wear sensitization coefficient, k*: is given by the ratio between the product between the volumetric wear and the friction coefficient and the mechanical work consumed by friction $[mm^3/J]$, etc. [18].

1.2.2 Analysis of the evolution of brake discs

The first brake discs were used in the 1890s in England. Frederick William Lanchester patented the first brake discs in Birmingham, England, in 1902, used successfully on Lanchester cars. However, in the first half of the 20th century, the only material from which the brake discs were made was copper. The performance of the braking discs made of copper was low.

Modern and high-performance brake discs appeared only in 1953, being also produced in England by Dunlop and used on the Jaguar C-Type race car. In 1955 the Citroen DS was the first French production car using modern braking discs, meantime the first production car with reliable brake discs appears only in England in 1956: Triumph TR3. The first production car with all-wheel brake discs was the Austin-Healey 100S. Compared to drum brakes, brake discs offer high performance due to more efficient cooling. As a result, they are much less exposed to the brake fade phenomenon. Also, brakes equipped with brake discs are much easier to recover [19].

Currently, the vast majority of cars are equipped with brake discs on the front axle. Some cheaper cars still keep the drum brake solution on the rear axle for cost reasons. This can also be a reasonable compromise, as the braking force on the rear is much lower than on the front axle. Based on the new requirements, researchers need to develop new materials and technologies for different existing applications. The field of friction materials is a sector where researchers must discover materials for new applications in increasingly severe, demanding working conditions. Friction and wear are the main processes that occur during the use of friction materials. Their best known applications are found in clutches and braking systems in cars, planes, space shuttles [20].

Ceramic discs are used in some high performance machines and heavy vehicles. The first ceramic braking project in the modern era of means of transport was carried out by British engineers working in the railroad industry for TGV (high speed train) applications in 1988. The aim was to reduce the weight, the number of brakes per axle and provide stable friction at high speeds and at all temperatures achieved by mechanical stress. The result was a carbon-reinforced ceramic product, which is now used in various forms of automobiles, railways but also applications for aircraft braking [21].

Due to the high heat tolerance and mechanical strength of composite ceramic discs, they are often used in exotic vehicles where the cost is not prohibitive for application. They are also found in industrial applications, where the weight-related properties of the ceramic disc and easy maintenance justify the relatively high cost compared to its metal alternatives. Composite brakes can withstand temperatures that would destroy flexible steel discs.

Composite ceramic brakes designed and manufactured for Porsche (PCCB) are made of carbon fiber silicone having a very high temperature resistance and benefit from a 50% reduction in weight compared to iron-carbon discs (therefore, the reduction of the unsuspended weight of the vehicle), a significant reduction in dust generation, intervals of a substantial increase in maintenance periods and increased durability in corrosive environments compared to conventional iron discs. This type of brake disc has been introduced on some of the more expensive models in the Porsche range in which the discs are ventilated internally as well as cast iron and have surface changes to increase temperature resistance properties [22].

1.2.3 Damage of the braking systems

The disks are usually damaged in one of four ways: scratches, micro-cracking, deformation and excessive oxidation. Car services will respond as much as possible to any problem with the discs by completely changing the discs. This is mainly done if the cost of a new disk can be less than the cost of labor to repair the old disk. From a mechanical perspective this is not necessary if the discs have reached the minimum recommended thickness of the manufacturer. If this thickness imposed by standards is reached, the use will become unsafe and there will be a risk of exposure to severe oxidation (especially ventilated discs) [23].

Automotive manufacturers recommend swivel brake discs as a general solution to problems related to vibration and brake noise. The machining process is performed on a lathe that removes a thin layer from the surface of the disc to clean minor damage and to achieve a uniform coarse restoration.

The identification and measurement of this phenomenon is performed using a dial indicator used on a fixed rigid base with the tip perpendicular to the front of the brake disc. The target measuring area is about 12.7 mm from the outer edge of the disc. The disk will be analyzed considering its rotation process. The difference between the minimum and maximum value on the dial is called the side beat. In the typical specifications for assembly stroke for passenger vehicles is around 0.0020 inches [24]. Excessive rolling is caused either by deformation of the disc or rolling in the wheel hub adjacent to the face or by contamination between the surface of the disc and the mounting surface of the preexisting hub. Determining the root cause of the indicator (lateral deviation) displacement requires disassembling the center disk [25]. Excessive rolling of the disc due to excessive rolling of the hub or contamination will have a period of at least 1 for each rotation of the brake disc.

The brake discs can be machined to eliminate the variation in thickness and excessive lateral rolling which is manifested by a deformation of the element in the braking system during its actuation. Processing can be done in situ (on-machine) or separately on a mechanical processing equipment. Both methods will eliminate the variation in the thickness of the brake discs. Mechanical processing of these elements from the braking system on machines with appropriate equipment can also eliminate excessive lateral rolling due to non-perpendicular contact made on the hub [26].

Improper mounting can distort (lead to scrap) the brake discs, as the screws securing the disc (or the wheel mounting system if the disc is simply attached to it, a system found on many cars), must be tightened progressive and uniform. The use of pneumatic tools for securing ring nuts is an extremely disadvantageous practical technique and a torque system can also be used. Depending on the system adopted, the vehicle manual will indicate a suitable model for tightening, as well as an assessment of the fastening system of the bolts as they should not be tightened in a circular system as some vehicles are sensitive to the force applied to the tightening bolts. The operation must be performed using a torque wrench.

Often the uneven transfer between the brake pad and the brake disc is the main cause for disc deformation [27]. In reality, most brake discs, which are diagnosed as deformed - scrap are simply a product of the uneven transfer of friction material of the brake pad. Uneven transfer of material from the brake pad to the disc will lead to a variation in disc thickness. When the thick section of the disc passes between the two brake pads, they will come off unevenly and the pressure in the brake pedal will increase slightly, known as the pedal pulsation [28]. The variation in brake disc thickness that can be felt by the driver is when it approaches approximately 0.17 mm (0.0067 inches) or larger (on all types of discs used in cars). The variation in the thickness of the brake disc has several causes, but three main mechanisms contribute most to the propagation of variations in the thickness of the disc related to the transfer of material from the brake pad, unevenly.

The first mechanism is related to the improper selection of brake pads for a particular application in terms of the materials chosen, the size and condition of their surface. Buffers (brake pads) that are effective at medium and low temperatures, such as during braking for the first time in cold weather, are made of materials that decompose unevenly at higher temperatures. These uneven decomposition results in an uneven deposition of material on the disk.

Another mechanism for the transfer of uneven material that settles on the brake disc is the improper pause in a combination of the brake pad and the brake disc. To obtain an adequate break time, the surface of the disc must be rectified (either by processing the contact surface or by replacing the disc as a whole) each time the plates are modified on a vehicle (activity that occurs every 2-3 years). Once this process is done, the brakes can be applied at maximum capacity several times in a row. This thing creates a less rough interface between the pad and the brake disc. When this is not done properly for the brake pads you will see an uneven distribution of stress and heat, resulting in a seemingly random, uneven deposition of friction material [29].

The third primary mechanism involved in the unequal transfer of material from the brake pad is known as pad printing. This happens when the brake pads heat up to the point where the material starts to break at the bottom and a transfer of the matter to the disc occurs. In the event of a corresponding rupture in the braking system (with correctly selected buffers), this transfer is natural and is a major contribution to the braking force generated by the disc brake pads. However, if the vehicle stops and the driver continues to apply the brakes, the pads will deposit a layer of material on the brake disc in the form of a brake pad. This variation of very small thickness can be the beginning of the uneven anti-friction material transfer cycle on the disc.

Once the disc has a certain level of variation in thickness, the deposition of the material to increase the wear coefficient unevenly will accelerate, sometimes leading to changes in the crystal structure of the material that makes up the disc in extreme situations [30].

As the brakes are applied, the pads slide on the surface of the disc differently. During the operation, the pads are forced outwards to pass through the thicker part of the disc. The force applied by the driver's foot on the brake pedal will naturally withstand this change

Materials Research Forum LLC
https://doi.org/10.21741/9781644901458

and create a higher pressure applied to the pads. Therefore, the passage of thicker sections will result in higher stress levels. This thing causes uneven heating of the disc surface, which involves two major problems. As the brake disc heats up unevenly, it will also expand unevenly. The thick sections of the disc will expand over the thin sections due to excess heat, so the difference in disc thickness is amplified. Also, due to the still uneven distribution of heat, an uneven transfer of the material on the brake pads will be achieved [31]. It will turn out that the brake disc is thicker on certain warmer sections that can take up the material on the brake pads even more than on the thinner cooler sections contributing to a further increase in the variation of the disc thickness.

In extreme situations this uneven heating can cause the disc material to crystallize leading to a phase change of the material. When the hot sections of the discs reach extremely high temperatures (1200 - 1300 °F or 649 - 704 °C) some percent of the carbon in the cast iron material will react with the iron molecules to form a carbide known as cementite. This iron carbide behaves very differently from the rest of the disc material of which it is composed because it is extremely heavy, hard and brittle and does not absorb heat well. Once the cementite is formed the integrity of the disc is completely compromised. Even if the surface of the disc is mechanically machined the cementite in the disc will not carry or absorb heat at the same rate as the surrounding cast iron, causing uneven disc thickness and uneven heating characteristics of the disc for heat treatment return [32].

Increased surface roughness occurs if the brake pads are not replaced promptly when they reach the end of their service life and are considered worn. After a sufficiently long period of friction of the material, the anti-friction steel plate for glued pads or the rivets for fixing the plates for riveted pads will support a direct contact on the wear surface of the disc that will decrease the braking power and scratches on the disc. Generally, a disc even if it is scratched /marked and has worked satisfactorily with existing brake pads, will be just as easy to use with new pads. If the marks on the disc are deeper but not excessive it can be repaired by processing by depositing a layer on the surface of the disc. This process can only be done a limited number of times because the disc has a minimum nominal thickness considered safe [33].

The minimum thickness is marked during manufacture on the disc, hub or on the edge of the disc. To prevent these marks on the surface of the brake disc, it is necessary to periodically inspect the brake pads for wear. When changing the tires it can be considered a logical moment of control of the brake pads and discs since this change must be made periodically according to the running time of the vehicle and all wheels must be changed allowing full visual access of the brake pads. Some types of alloy wheels and specific braking modes will provide enough free space to view the pads without removing the wheel. When the brake pads are close to the wear point, they must be replaced immediately because complete wear will lead to damage by increased roughness and increased brake insecurity [34].

Many of the brake pad systems include a soft steel spring for compressing the buffer assembly and which is designed to start sliding on the disc when the block of material is exhausted. The result is a loud metallic noise that serves to alert the vehicle user to notify him that he needs service and this situation will not normally materialize by damaging the disc if the brakes are repaired in real time. A new set of brake pads is considered for replacement if the thickness of the anti-friction material is the same or less than the thickness of the reference support steel (in Pennsylvania, USA the standard is 1/32 inch)

The occurrence of cracks in the brake discs is largely limited to the perforated discs due to the uneven expansion speed in severe disc drive environments. Manufacturers, who use drilled discs, do so for two reasons: appearance or as a function of reducing the weight of the unsuspended brake with the assumption that the new mass of the remaining brake will absorb the temperature during the race [35].

A brake disc is also a radiator in the braking system but the loss of mass of the radiator can be balanced by the increased surface area of radiation needed to remove excessive heat flow. Small capillary cracks can occur in any metal disc, mechanically machined in a cross, in the form of a normal wear mechanism, but if this occurs severely, the disc will give up with catastrophic consequences. No repair method for cracks is known, and if their cracking becomes severe, the disc needs to be replaced.

The discs are currently made of cast iron and a certain amount of what is known as surface rust is normal. The contact area between the brake disc and the brake pads will be kept clean by regular use. A vehicle that is stored for a long time develops areas with significant rust in the contact area and will reduce braking power for a time, until the rusty layer is removed again. Over time, ventilated brake discs can develop areas with severe rust that lead to advanced corrosion inside the vents that compromise the strength of the structure and lead to an acute need to completely replace the brake disc.

1.3 Materials, technologies and equipment used in the processes of obtaining brake discs

The main materials used to make discs in the braking systems and the problems they present in the last twenty years are identified. The main technologies and equipment's used to make metallic materials by casting are presented.

1.3.1   Classic friction materials

Fifteen years ago, automotive friction manufacturers used only two basic components: an asbestos-based component for drum brakes and a semi-metallic or asbestos-based component for brake discs. Today, their number has increased, with the addition of various organic (NAO - Non Asestos Organic) or semi-metallic components [38]. Research on friction materials has led to a constant relationship between the four most important characteristics:

1) vehicle weight;

2) active braking surface;

3) brake configuration;

4) weight and design of the rotor, a ratio that can indicate the degree of heating of the brake material and the time in which the heat obtained will be dissipated.

A material containing more than 30% by weight of Fe-C alloy is considered a semi-metallic friction component. In contrast, polymeric friction materials (asbestos-free) contain a small amount of iron. Recipes for friction materials contain fibers, fillers, binders and resins. The choice of materials for a given tribological application starts from the analysis of surface geometry, contact pressure, type of movement, relative sliding speed, nature and thickness of the material, as well as the atmosphere of the system (temperature, humidity and specific chemical reactions) [36]. The analysis of tribological phenomena requires a multidisciplinary approach combining mechanics and physics surface, and materials and interface chemistry. One of the basic rules used in the selection of materials for a tribological coupling is the so-called tribological compatibility rule according to which materials that are mutually insoluble have a low tendency to adhere, having a good coupling for friction materials. However, there are two reasons why this rule does not apply:

- solubility is an intrinsic property of massive materials while friction depends primarily on surface characteristics (hardness, crystal structure, surface energy, topography);

- the solubility criteria do not consider chemical transformations (oxidations, phase transformations, the appearance of new compounds) or structural damage (allotropic transformations, recrystallizations) that occur during friction and that can alter the surface properties of the materials in contact .

The braking system is an important component of cars. It consists of four subsystems, namely: the drive system; transmission system; wheel braking system and electronic safety system. The actuating system has the role of actuating and adjusting the braking force. It is transmitted to the wheels via the brake pipes and cables via the brake fluid. The wheel braking system ensures the actuation of the braking devices by pressing the pads and brake shoes onto the discs, respectively the brake drums. The brake discs are connected via a piping system to the main brake cylinder. The braking system as a whole contain also the parking brake, power brake, ABS, ECP, ESP, devices capable of improving its efficiency [37].

1.3.2 Friction materials with ceramic matrix

As polymer matrix composites and metal matrix composites cannot operate at temperatures above 1000 °C, research has been conducted in the field of ceramic matrix friction composites. Due to the high cost of both raw materials and technologies for developing these materials, most of their applications are found in the aerospace, aeronautics and nuclear industries. Among the most common ceramic matrix composite materials are silicon carbide and carbon matrix materials reinforced with carbon fibers (C/C-SiC).

Carbon fiber-reinforced silicon and carbon matrix composite materials have been developed as brake materials since the late 1990s [38]. These materials have a higher coefficient of friction than iron or carbon/carbon-based materials, as well as a high degree of wear and resistance to thermal shock. The value of the coefficient of friction determined for operation in dry environment was 0.38, while for operation in wet environment, it was 0.35. It can be seen that this parameter does not vary much between the two operating conditions, compared to the friction materials presented above. At a test cycle, in dry friction conditions, the value of the wear degree was 1.1 μm, respectively 0.7 μm in wet friction conditions [39]. Silicon carbide-based composites reinforced with carbon fibers have superior tribological properties to those of gray cast iron or carbon/carbon materials. These materials are suitable for advanced tribological applications due to their low density, high resistance to thermal shock and good wear resistance.

Improving the wear resistance of this class of friction materials can be achieved by ceramic coatings. The low density of carbon/carbon-silicon carbide materials reduces the weight of the braking systems by approximately 40 - 60% [40]. The high wear resistance causes an increase in the number of brakes to which a part made of these materials can be subjected. Another feature of these materials is the ability to absorb a large amount of kinetic energy which is subsequently converted into heat. Within the manufacturing process of some components from ceramic matrix materials as well as those from carbon-based materials, three stages can be distinguished. These stages are:
- formation of the fibrous skeleton with oriented fibers, arranged randomly;
- consolidation of the fibrous skeleton with the matrix material and
- mechanical processing of the product.

The consolidation of the fibrous skeleton can be done by solid, liquid and gaseous methods, or by combinations thereof. Ceramic friction materials are of growing interest to

researchers. As with other types of friction materials, the literature provides a wide range of information about these composites. The manufacturing process involves a set of operations in various equipment to make liquid alloys with the composition, degree of purity and prescribed temperature, by melting the solid load, overheating and processing it into a liquid state. The following are used as processing units in cast iron foundries:
- cupola;
- flame furnaces (fixed or rotary);
- electric furnaces with induction or spring heating.

In the case of making cast iron in the cupola, the load of the unit consists of two main parts: the coke of coal bed (at the bottom of the vat) and the part made of several load segments (these are located up to the loading hole). A load segment contains metallic materials at the bottom and above them combustion materials, for example coke and also a limestone load. Usually, in practice, the metal load is made of the following materials: first cast iron, cast iron scrap and steel. The main metallurgical treatment is the correction of the chemical composition that is made with ferroalloys during the refining of the metallic mass. To perform this step, batches of molten material are periodically taken and chemically analyzed with the spark spectrometer in order to control the chemical composition of the metal melting.

The elaboration of cast irons in furnaces with induction heating is performed and has the following advantages: obtaining a precise and homogeneous composition of the alloys, higher overheating temperatures and the possibility of using cheap loads (scrap metal, cast iron chips, waste, etc.).

In the case of induction heating, it is necessary to cut to size the metal load introduced in the furnace. The cutting operation depends on the size of the furnace and its melting capacity. The materials that make up the load are cleaned to remove grease and emulsions from the surface. As an additional operation for removing oxides from the charge, a reducing mixture is introduced into the furnace with the charge.

The thermal efficiency of the oven is closely related to the temperature of the load, having low values at melting and high at overheating. It is therefore necessary to put the load in the furnace after a preheating in which accumulates about 20 - 30% of the heat needed for melting. The optimum preheating temperature is between 430 - 650 °C. In the case of melting and overheating of cast irons, the heating time depends on the frequency of the current and the specific power used. For the elaboration of cast irons, you can also use hearth furnaces (heated with flame or electric) as well as combinations of several types of furnaces (duplex or triplex elaboration processes). When making cast irons in electric arc furnaces, the lining of the furnace is acidic and only when the basic removing of phosphorus (dephosphorylation) and desulfurization of the bath is sought.

In the case of arc furnaces, it is ensured to obtain high temperatures, thus achieving a higher melting efficiency and productivity than induction furnaces. Due to the high temperature of the slag, through it can be performed processes of refining, desulfurization, dephosphorylation, etc. The disadvantages of this method are related to a low efficiency of overheating (less than 20%); high release of dust and smoke during melting and advanced combustion of components.

1.4 Properties of friction alloys used in car braking systems

Friction is evaluated by numbers and letters that reflect the coefficient of friction. In this system the value 1 is the largest possible number and denotes a high level of friction. Most brake pads can reach this level of friction. The letter system has been developed to define typical friction values. These values are taken at a standard pressure depending on a standard surface of the material. Table 1.1 shows the friction coefficients and the corresponding letter in the code.

Table 1.1. Code letters and corresponding friction values

Coding letter	Coefficient of friction
C	0-0.15
D	0.15-0.25
E	0.25-0.35
F	0.35-0.45
G	0.45-0.55
H	0.55 and >

It is desirable to use a material with the highest possible coefficient of friction for all vehicles. While this situation is perfect for braking, there are downsides. If a material with a high coefficient of friction is used, the brake disc or drum will be damaged more quickly by wear. To this end, a vehicle must be equipped with its own braking system, one that does not prematurely wear out the other components of the system. The mechanical energy of motion that the rubbing materials absorb must be converted into another form of energy. In a braking system the energy of motion (kinetic energy) is converted into heat. The law of conservation of energy is known from physics. This law postulates that energy cannot be destroyed but can be transformed from one form to another. In a braking system the kinetic energy of a vehicle is converted into a large amount of heat by the braking system.

The brake system bearings will generate a high amount of heat. This process can be seen in race cars during video broadcasts. In some cases, you can even see the brake discs with a color change to orange-red, under the influence of temperature. This case is an example of excessive use of the braking system. These car systems are equipped with cold air ducts to the brake discs and braking systems to support them for several laps.

Many vehicles initially used braking systems on the 4-wheel drums. In many cars of that time, the drums were supported on all four wheels. Larger cars used ribbed drums. However, these drum brake systems have some specific problems. First, they retain water, which decreases in capacity and properties of action during rainstorms or after passing through a puddle. Second, they dissipate the heat and they will give in to long or steep descents or after repeated sudden stops. Lastly, their braking distance is much longer than that characteristic of disc brakes. Disc brakes have been developed to eliminate these problems with drum brake systems. The concept of disc brakes is not difficult to understand. Similar to systems applied to stop bicycles, instead of stopping the wheel with a brake drum on the axle, a tightening on both sides of the rim is used to stop it.

Brake discs are made of various materials processed to low tolerances. There are single or double disc brake systems. Disc-based braking systems have several advantages over the drum system. First, the contact surface between the metal that is in contact with the brake material during a rotational movement is approximately 30 percent larger. Second, the brake disc is located in the airflow behind the tire. This position allows the brake disc and brake caliper to cool more efficiently than the drum brake system, and the cooling process will prevent wear and tear of the brake materials. Moreover, no disc brake system will be as much affected by dirt or mud as the one on the drum, because the pads are located close to the brake disc and will constantly clean the disc.

The disc brake system allows the vehicle to stop in a straight line, even if part of it has passed or is in a watery or damaged area. The centrifugal force also has the role of cleaning the brake disc by removing impurities from it, thus keeping the brake system clean.

References

[1]   M. R. Ishak, A. R. A Bakar, A. Belhocine, J. M. Taib, W. Z. W. Omar, Brake torque analysis of fully mechanical parking brake system: Theoretical and experimental approach. Measurement. 94, (2016) 487-497. https://doi.org/10.1016/j.measurement.2016.08.026

[2]   M. H. Westbrook, The electric car: development and future of battery, hybrid and fuel-cell cars. London, Institution of Electrical Engineers, 2001. https://doi.org/10.1049/PBPO038E

[3]   A. J. Day, T.P. Newcomb, The Dissipation of Frictional Energy From the Interface of an Annular Disk Brake. Proceedings Institute of Mechanical Engineers. 198 (1984) 201-209. https://doi.org/10.1243/PIME_PROC_1984_198_146_02

[4]   A. K. Baker, Vehicle braking. Pentech Press, London, 1986

[5]   R. Limpert, Brake design and safety. Warrendale, Pa., Society of Automotive Engineers, 1999. https://doi.org/10.4271/9780768027105

[6]   J. E. Hunter, S. S. Cartier, Brake Fluid Vaporization as a Contributing Factor in Motor Vehicle Collisions. International Congress and Exposition, Detroit, Michigan, SAE, 1998. https://doi.org/10.4271/980371

[7]   T. K. Kao, J. W. Richmond, et al. Brake disc hot spotting and thermal judder: an experimental and finite element study. Int. J. Vehicle Des. 23(3/4), (2000) 276-296. https://doi.org/10.1504/IJVD.2000.001896

[8]   A. Jerhamre, C. Bergstrom, Numerical Study of Brake Disc Cooling Accounting for Both Aerodynamic Drag Force and Cooling Efficiency. SAE 2001 World Congress, Detroit, Michigan, 2001. https://doi.org/10.4271/2001-01-0948

[9]   C. Zhang, L. Zhang, Q. Zeng, S. Fan, L. Cheng, Simulated three-dimensional transient temperature field during aircraft braking for C/SiC composite brake disc. Mater Design. 32 (2011) 2590 – 2595. https://doi.org/10.1016/j.matdes.2011.01.041

[10]   M. Kubota, T. Hamabe, et al., Development of a lightweight brake disc rotor: a design approach for achieving an optimum thermal, vibration and weight balance. JSAE Review 21 (2000) 349-355. https://doi.org/10.1016/S0389-4304(00)00050-3

[11]   J. Takadoum, Materials and surface engineering in tribology, British Library, 2007. https://doi.org/10.1002/9780470611524

[12]   ASM Handbook. Friction, lubrication and wear technology, vol. 18, ASM. International The Materials Information Company, United States of America, 1997.

[13]   H. Czichos, Tribology - a systems approach to the science and technology of friction lubrication and wear, Elsevier Sci. Pub. Co., Amsterdam, 1978. https://doi.org/10.1016/0301-679X(78)90209-8

Materials Research Forum LLC
https://doi.org/10.21741/9781644901458

[14]   D. Pavelescu, M. Muşat, A. Tudor, Tribology, Ed. Didactică şi Pedagogică, Bucureşti, 1977

[15]   T. Ghrib, New tribological ways, Intech, Rijeka, 2011. https://doi.org/10.5772/637

[16]   G. W. Stachowiak, A. W. Batchelor, Engineering tribology – Third edition, British Library, 2005.

[17]   P. S. M. Jena, J. K. Sahu, R. K. Rai, S. K. Das, R. K. Singh, Influence of duplex ferritic-austenitic matrix on two body abrasive wear behaviour of high chromium white cast iron. Wear. 406–407 (2018) 140-148. https://doi.org/10.1016/j.wear.2018.04.004

[18]   G. E. Totten, ASM HANDBOOK, Friction, Lubrication, and Wear Technology, ASM International, Volume 18, 2017. https://doi.org/10.31399/asm.hb.v18.9781627081924

[19]   S. M. Savaresi, M. Tanelli, Active Braking Control Systems Design for Vehicles, Springer, 2010. https://doi.org/10.1007/978-1-84996-350-3

[20]   K. Saw, S. Shankar, S. Chattopadhyaya, P. Vilaca, Microstructure Evaluation of Different Materials after Friction Surfacing - A Review. Materials Today: Proceedings. 5, (2018) 24094-24103. https://doi.org/10.1016/j.matpr.2018.10.203

[21]   S. Zhu, J. Cheng, Z. Qiao, J. Yang, High temperature solid-lubricating materials: A review. Tribol. Int. 133 (2019) 206-223. https://doi.org/10.1016/j.triboint.2018.12.037

[22]   M. Haghshenas, A. P. Gerlich, Joining of automotive sheet materials by friction-based welding methods: A review. Engineering Science and Technology, an International Journal, 21 (2018) 130-148. https://doi.org/10.1016/j.jestch.2018.02.008

[23]   C.G. He, Y. Z. Chen,Y. B. Huang, Q. Y. Liu, W. J. Wang,. On the surface scratch and thermal fatigue damage of wheel material under different braking speed conditions. Eng. Fail. Anal. 79 (2017) 889-901. https://doi.org/10.1016/j.engfailanal.2017.06.017

[24]   J. Chen, J. Yu, K. Zhang, M. Yan, Control of regenerative braking systems for four-wheel-independently-actuated electric vehicles. Mechatronics. 50 (2018) 394-401. https://doi.org/10.1016/j.mechatronics.2017.06.005

[25]   W. J. Wang, F. Wang, K. K. Gu, H. H. Ding, M. H. Zhu, Investigation on braking tribological properties of metro brake shoe materials. Wear. 330–331 (2015) 498-506. https://doi.org/10.1016/j.wear.2015.01.057

[26] C. Wang, W. Zhao, W. Li, Braking sense consistency strategy of electro-hydraulic composite braking system. Mech. Syst. Signal Pr. 109 (2018) 196-219. https://doi.org/10.1016/j.ymssp.2018.02.047

[27] R.Gadow, A. Kienzle, Processing and manufacturing of C-fiber reinforced SiC composites for disk brakes, Proceedings of the 6th International Symposium on Ceramic Materials and Components for Engines, (1997) 412-418.

[28] S. Pal, D. Mallikarjuna Reddy, S. Saha, A. Sharma, Design and develop a novel brake lighting mechanism for intensity of braking: automobile applications, Materials Today: Proceedings, 5 (2018) 13069-13078. https://doi.org/10.1016/j.matpr.2018.02.294

[29] E. Davin, A.-L.Cristol, J.-F. Brunel, Y. Desplanques, Wear mechanisms alteration at braking interface through atmosphere modification, Wear 426–427 (2019) 1094-1101. https://doi.org/10.1016/j.wear.2019.01.057

[30] M. Kchaou, A. R. Lazim, M. K. Abdul Hamid, A. R. Abu Bakar, Experimental studies of friction-induced brake squeal: Influence of environmental sand particles in the interface brake pad-disc. Tribol. Int. 110 (2017) 307-317. https://doi.org/10.1016/j.triboint.2017.02.032

[31] J.-G. Bauzin, N. Keruzore, N. Laraqi, A. Gapin, J.-F. Diebold, Identification of the heat flux generated by friction in an aircraft braking system, Int. J. Therm. Sci. 130 (2018)449-456. https://doi.org/10.1016/j.ijthermalsci.2018.05.008

[32] S. Abbasi, S. Teimourimanesh, T. Vernersson, U. Sellgren, R. Lundén,. Temperature and thermoelastic instability at tread braking using cast iron friction material, Wear. 314, 1–2 (2014) 171-180. https://doi.org/10.1016/j.wear.2013.11.028

[33] A. R. Zulhishamuddin, S. N. Aqida, M. Mohd Rashidi, A comparative study on wear behaviour of Cr/Mo surface modified grey cast iron. Optics & Laser Technology. 104 (2018) 164-169. https://doi.org/10.1016/j.optlastec.2018.02.027

[34] P. Shiva, A review on properties of conventional and metal matrix composite materials in manufacturing of disc brake. Materials Today: Proceedings. 5, (2018) 5864-5869. https://doi.org/10.1016/j.matpr.2017.12.184

[35] M. Pevec, G. Oder, I. Potrč, M. Šraml, Elevated temperature low cycle fatigue of grey cast iron used for automotive brake discs. Eng. Fail. Anal. 42, (2014) 221-230. https://doi.org/10.1016/j.engfailanal.2014.03.021

[36]   L. Zhuan, X. Peng, X. Xiang, Z. Su-Hua, Tribological characteristics of C/CsiC braking composites under dry and wet conditions. T. Nonferr. Metal. Soc. 18, (2008) 1071-1075. https://doi.org/10.1016/S1003-6326(08)60183-1

[37]   A. A. Yevtushenko, M. Kuciej, O. Yevtushenko, Modelling of the frictional heating in brake system with thermal resistance on a contact surface and convective cooling on a free surface of a pad. International Journal of Heat and Mass Transfer. 81 (2015) 915-923. https://doi.org/10.1016/j.ijheatmasstransfer.2014.11.014

[38]   J. Wang, M. Lin, Z. Xu, Y. Zhang, Z. Shi, J. Qian, G. Qiao, Z. Jin, Microstructure and mechanical properties of C/C-SiC composites fabricated by a rapid processing method. J. Eur. Ceram. Soc. 29 (2009) 3091 – 3097. https://doi.org/10.1016/j.jeurceramsoc.2009.04.036

[39]   W. Krenkel, B. Heidenreich, R. Renz, C/C-SiC composites for advanced friction systems. Adv. Eng. Mater. 4 (2002) 427-436. https://doi.org/10.1002/1527-2648(20020717)4:7<427::AID-ADEM427>3.0.CO;2-C

[40]   A.P. Garshin, V.I. Kulik, A.S. Nilov,. A new generation of constructional materials (review) - Braking friction materials based on fiber-reinforced composites with carbon and ceramic matrices. Refractories and Ind. Ceramics. 49, (2008) 391 – 396. https://doi.org/10.1007/s11148-009-9099-6

Chapter 2

Corrosion Dynamics in Fe-C Alloy Systems

C.D. Florea[1], C. Bejinariu[1], R. Cimpoesu[1], N. Cimpoesu[1]*

[1]Faculty of Materials Science and Engineering, "Gheorghe Asachi" Technical University of Iasi, Romania

nicanor.cimpoesu@tuiasi.ro

Abstract

A mathematical model is constructed based on the chaotic system-complex system transition in the description of corrosion resistance dynamics in Fe-C (cast iron) alloy systems. In such a context, fractality/multifractality gives the system archeology/histories, situations in which a logistic type fractal / multifractal law will operate in the description of corrosion resistance dynamics in Fe-C (cast iron) alloy systems. The results of the theoretical model can explain cyclic diagrams and corrosion tafel diagrams in Fe-C (cast iron) alloy systems.

Keywords

Mathematical Model, Corrosion, Fractality, Cast Iron

2.1 Nonlinearity and chaos in complex systems such as Fe-C alloys

Determinism in analysis of dynamics in materials science does not necessarily involve either regular behavior (periodic movements, self-structures. etc.) or predictability in the behavior of complex systems such as Fe-C (cast) alloy systems [1,2].

In linear analysis, on which the science of 20th century materials was based exclusively, unlimited predictability was an automatic quality of the dynamics of complex systems such as Fe-C alloys (cast iron). The development of nonlinear analysis and the discovery of laws governing chaos have shown not only that the reductionist method of dynamic analysis on which all materials science has so far been based has limited applicability but also that infinite predictability is not an attribute of complex Fe- C type systems, but a natural consequence of their simplification by linear treatment [3,4].

Consequently, nonlinearity and chaos specify common behaviours, universality in the laws that dictate the dynamics in complex systems such as Fe-C alloys (cast iron).

Nonlinearity and chaoticity for the complex system such as Fe-C alloys (cast iron) are both structural and functional, the interactions between its entities (also called structural units) causing mutual conditioning microscopically-macroscopically, locally-globally, individually-collectively, etc. [3,4].

In such a framework, the universality of the legitimacy of the dynamics of complex systems such as Fe-C alloys (cast iron) becomes natural/obvious and must be reflected in the mathematical procedures used. Some authors in the field of materials engineering are increasingly discussing holographic implementations in describing the dynamics of complex systems such as Fe-C alloys (cast iron) [5,6].

Habitually, the current theoretical models used in describing the dynamics of complex systems such as Fe-C alloys (cast iron) are based on the assumption, still unjustified, of the differentiability of the variables that describe it. The success of these models must be understood gradually/sequentially, in areas where differentiability and integrability are still valid. Differentiable and integrable mathematical procedures suffer when we want to describe evolutions in the dynamics of complex systems such as Fe-C alloys (cast iron) because only they involve nonlinearity and chaoticity [5,6].

To describe the dynamics of complex systems such as Fe-C alloys (cast iron), while remaining dependent on differentiable and integrable mathematical procedures, it is necessary to explicitly introduce scale resolution in the expressions of the variables that describe them and implicitly in the expressions of the fundamental equations governing these evolutions in dynamics of complex systems such as Fe-C alloys (cast iron) [7-10].

This means that any variable, dependent in the classical sense, both on spatial and temporal coordinates, depends in the new mathematical structure on the scale resolution. That said, instead of operating, for example, with a single variable described by a strictly non-differential mathematical function, we will operate only with approximations of this mathematical function obtained by mediating it at different scale resolutions. Consequently, any variable meant to describe dynamics in complex systems such as Fe-C alloys (cast iron) will function as the limit of a family of mathematical functions, which is indistinguishable for a zero-scale resolution and differentiable for a non-zero-scale resolution [7-10]. This way of describing the dynamics of the complex system such as Fe-C alloys (cast iron), where any measurement is made at finite scale resolutions, obviously involves both the development of new geometric structures and theories consistent with these geometric structures for which the laws of motion, invariant to spatial and temporal transformations are "integrated" with laws of scale invariant to transformations of scale resolutions. In our opinion, such a geometric structure can be based on the concept of fractal/multifractal and the corresponding theoretical model Theory of Relativity of Scale [9-12].

2.2 Chaotic system-complex system transition for Fe C type alloys by fractality/multifractality - the mathematical model

As mentioned above, Fe-C (cast iron) alloys are complex systems. The complexity of Fe-C systems refers both to the collective behaviour of Fe-C systems (dictated by the extremely large number of nonlinear interactions between structural units/entities) and to the constraints that the Fe-C system bears in relation to the environment. From such a perspective, the Fe-C system evolves far from the state of equilibrium (at the edge of chaos, between determinism and chance), in a critical state constructed from an archeology/history of "unpredictable and unexpected events" through feedback cycles, self-structuring, etc. Thus, archeology/history is substantiated as the main characteristic of the complex system such as Fe-C alloys (cast iron).

However, the absence of archeology/history is specific to chaotic systems. Here the behaviour of any such Fe-C system is dictated by the relatively small number of nonlinear interactions between its structural units and has as its purpose the "order" of chaos. In a sense, the chaotic Fe-C system can be assimilated to a subset of complex systems whose archeology/history has been suspended. In such a context, is it still possible for chaotic systems (systems without archeology/history) to mimic complex systems (systems with archeology/history)? The answer is affirmative only if the chaotic systems are attributed to archaeologies/histories. In this regard, let us first admit that Fe-C type alloys (cast iron) as chaotic systems can withstand corrosion resistance dynamics based on logistic law:

$$\frac{dJ}{dV} = RJ\left(1 - \frac{J}{K}\right) \tag{2.1}$$

with $R > 0, K > 0, J(0) = J_0$ $\tag{2.2}$

where: J is the current density below the potential V;

  J_0 current density below zero potential;

  R and K   structure constants specific to corrosion resistance dynamics

The solution of equation (2.1):

$$J(V) = \frac{J_0 K}{J_0 + exp(-RV)(K - J_0)} \tag{2.3}$$

can describe for different values of R, K and J_0 complex dynamics of corrosion resistance for various potentials, Fig. 2.1-2.4.

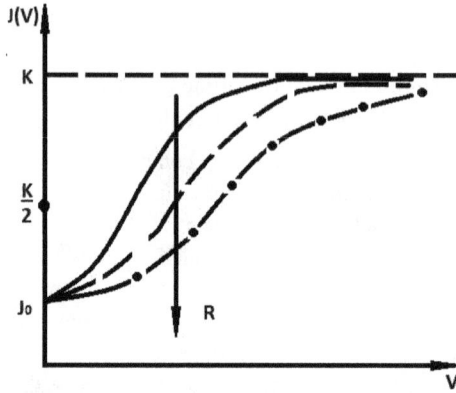

Figure 2.1. Theoretical corrosion resistance curves with the same $K > J_0$ and R differently described by the solution of the logistic type equation (2.1)

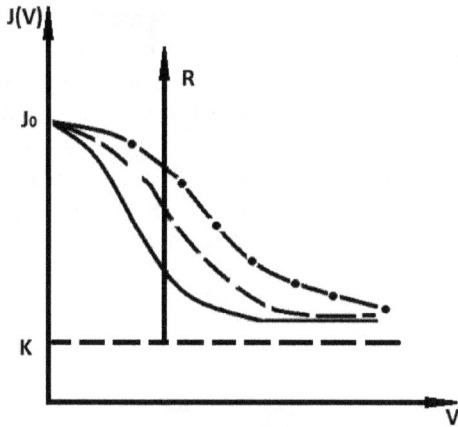

Figure 2.2. Theoretical corrosion resistance curves with the same $K < J_0$ and R differently described by the solution of the logistic type equation (2.1)

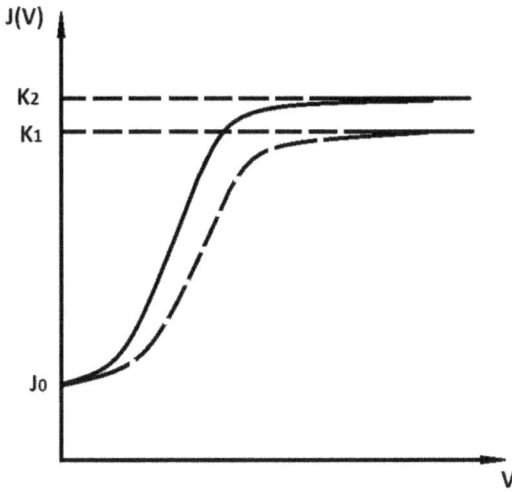

Figure 2.3. Theoretical corrosion resistance curves with the same R and K different and K> $_{J0}$ described by the solution of the logistic type equation (2.1)

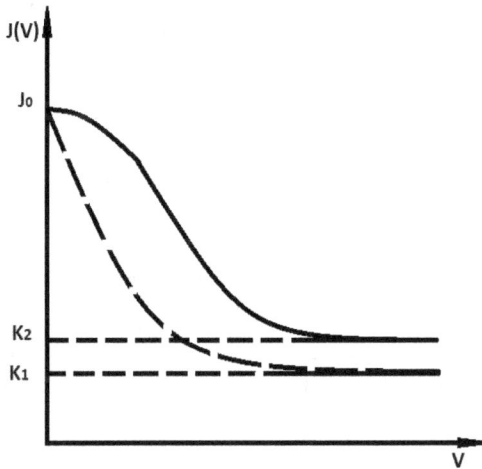

Figure 2.4. Theoretical corrosion resistance curves with the same different R and K and K <J_0 described by the solution of the logistic type equation (2.1)

A possible procedure for assigning archaeologies/histories of the Fe-C alloy, a situation in which it changes its status, from chaotic to complex system, is through fractality/multifractality, to the corrosion resistance curves of Fe-C alloys [12]. Then both the variables describing dynamics of corrosion resistance and constants of structures specific to corrosion mechanisms become dependent on the scale resolution in the form:

$$J = J(V, \mu, F(\alpha)), R = R(\mu, F(\alpha)), K = K(\mu, F(\alpha)), J(0, \mu, F(\alpha)) = J_0(\mu, F(\alpha)),$$

$$\alpha = \alpha(D_F) \tag{2.4}$$

Where

J	is the multifractal current density below the non-zero multifractal potential;
J_0-	multifractal type current density below zero multifractal type potential;
R și K	are the structural constants specific to the multifractal corrosion resistance mechanisms, with the multifractal scale resolution;
F=F(α)	singularity spectrum of singularity index;
α	functionally dependent index of the fractal dimension DF of the corrosion curve

The singularity spectrum F (α) will allow the identification of some universality classes in the field of dynamic Fe-C alloy type systems with corrosion resistance.

In such a context, the relations (2.1) - (2.3) maintain their functionality, with the only observation that in this case the quantities we operate with (dynamic variables, structure coefficients, etc.) are given by fractal/multifractal functions.

2.3 Validation of the theoretical model

The above considerations have some obvious consequences, presented below.

Between two points in the fractal/multifractal space I-V there are an infinity of fractal/multifractal curves. In particular, if the fractal/non-fractal curves are elliptical, according to a fundamental theorem in the theory of elliptic functions [13], two elliptical curves (fractal/non-fractal) of the same length cannot be superimposed, Fig. 2.5.

Global scale corrosion dynamics are expressed as overlaps of local scale corrosion dynamics based on the principle of superposition of scale resolutions in the fractal theory of motion [11]. Let two characteristics I = I (V) be at two different local scale resolutions, Fig. 2.6a,b.

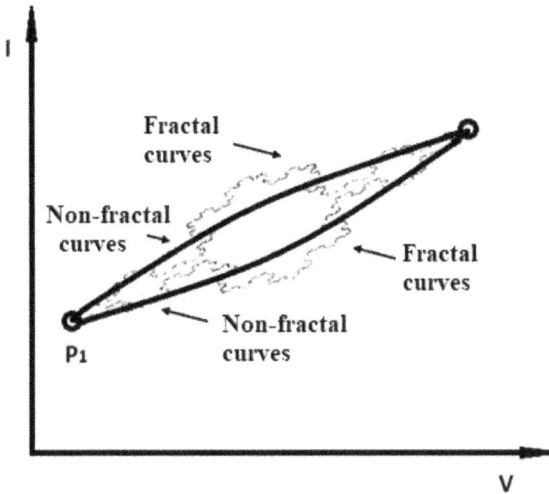

Figure 2.5. Families of elliptical curves (fractal and non-fractal) with corrosion resistance, of the same length but which can overlap.

Now the functionality of the principle of superposition of scale resolutions allows to obtain a theoretical diagram which, at least qualitatively, can be put in correspondence with the experimental Tafel diagrams [14-20].

Correspondence between the theoretical model and the experimental data can be obtained based on the following specifications:

i) The fact that between two points of a multifractal space there are an infinity of fractal / multifractal curves and in particular elliptic curves with the property that they have the same length but they cannot be superimposed, shows that both by a convenient choice of points in the fractal / multifractal space as well as choosing sufficient number of sizes I and V we can construct theoretical curves that can reproduce quite accurately the experimental cyclic corrosion resistance diagrams presented in Fig. 2.7.

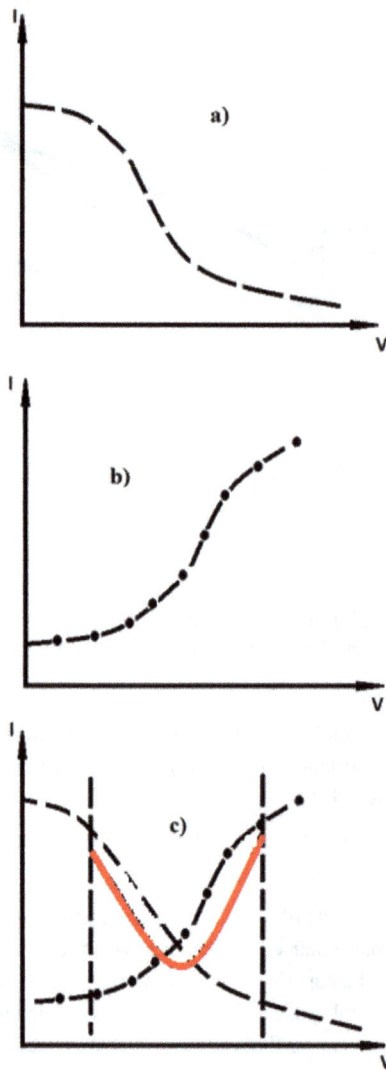

Figure 2.6. Corrosion resistance curves at different scale resolutions (a, b) respectively at global scale resolutions (c).

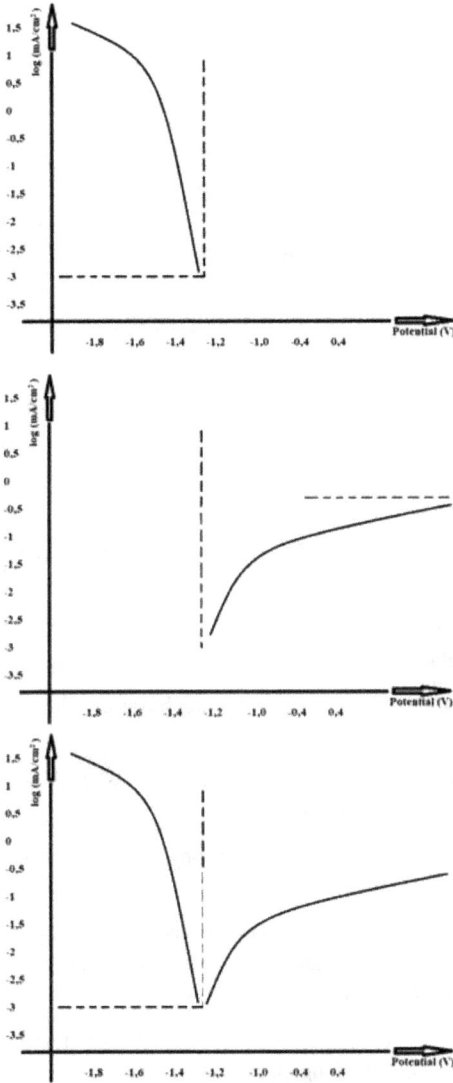

Figure 2.7 Sequences at various scale reports for reproducing Tafel experimental diagrams based on the theoretical model.

ii) The functionality of the stair overlay principle also allows both by an adequate selection of stair distributions (either local or fractal) and by a convenient choice of sizes related to the Tafel experimental graph in Fig. 2.7. For example, we can choose the ratio of real scales, for a general one see Fig. 2.7.

References

[1]   R. Badii, A. Politi, Complexity: Hierarchical structures and scaling in physics. Cambridge University Press, New York, 1997. https://doi.org/10.1017/CBO9780511524691

[2]   M. Mitchell, Complexity: A Guided Tour, Oxford University Press. New York, 2009.

[3]   P.C. Cristescu, Nonlinear dynamics and chaos: fundamentally theoretical and applications , Bucharest , Romanian Academy Press. 2008.

[4]   E.A. Jackson, Perspectives of Nonlinear Dynamics. Vol. 142, , Cambridge University Press, New York: 1993

[5]   O.D Michel, B. G. Thomas, Mathematical Modeling for Complex Fluids and Flows, Springer, New York, 2012.

[6]   Y.H. Thomas, Multi-Scale Phenomena in Complex Fluids: Modeling, Analysis and Numerical Simulations, World Scientific Publishing, 2009.

[7]   B.B. Mandelbrot, Les objets fractals: Forme, hazard, et dimension, Flammarion. , Paris: 1972.

[8]   B.B. Mandelbrot, The Fractal geometry of nature, W. H. Freeman New York, 1983. https://doi.org/10.1119/1.13295

[9]   L. Nottale, Fractal Space-Time and Microphysics: Towards A Theory Of Scale Relativity, World Scientific, Singapore, 1993. https://doi.org/10.1142/1579

[10] L. Nottale, Scale Relativity and Fractal Space-Time: A New Approach to Unifying Relativity and Quantum Mechanics. Imperial College Press, London, 2011. https://doi.org/10.1142/p752

[11] I. Merches, M. Agop, Differentiability and Fractality in Dynamics of Physical Systems, World Scientific, Singapore, 2013.

[12] J. Cresson, Scale calculus and the Schrödinger equation. J. of Mathem. Phys. 44 (2003) 4907. https://doi.org/10.1063/1.1618923

[13] J.V. Armitage, W.F. Eberlein, Elliptic Functions. Cambridge University Press. New York, 2009.

[14] B. Istrate, C. Munteanu, R. Cimpoesu, N. Cimpoesu, O.D. Popescu, M. D. Vlad, Microstructural, Electrochemical and In Vitro Analysis of Mg-0.5Ca-xGd Biodegradable Alloys, Applied Sciences-Basel 11 (2021) 981. https://doi.org/10.3390/app11030981

[15] C. Bejinariu, D.P. Burduhos-Nergis, N. Cimpoesu, Immersion Behavior of Carbon Steel, Phosphate Carbon Steel and Phosphate and Painted Carbon Steel in Saltwater, Materials 14 (2021) 188. https://doi.org/10.3390/ma14010188

[16] N. Cimpoesu, F. Sandulache, B. Istrate, R. Cimpoesu, G. Zegan, Electrochemical Behavior of Biodegradable FeMnSi-MgCa Alloy, Metals 8 (2018) 541. https://doi.org/10.3390/met8070541

[17] B. Istrate, J.V. Rau, C. Munteanu, I.V. Antoniac, V. Saceleanu, Properties and in vitro assessment of ZrO2-based coatings obtained by atmospheric plasma jet spraying on biodegradable Mg-Ca and Mg-Ca-Zr alloys, Ceramics international 46 (2020) 15897-15906. https://doi.org/10.1016/j.ceramint.2020.03.138

[18] M.G. Zaharia, S. Stanciu, R. Cimpoesu, I. Ionita, N. Cimpoesu, Preliminary results on effect of H2S on P265GH commercial material for natural gases and petroleum transportation, Applied surface science 438 (2018) 20-32. https://doi.org/10.1016/j.apsusc.2017.10.093

[19] J. Izquierdo, G. Bolat, N. Cimpoesu, L.C. Trinca, D. Mareci, R.M. Souto, Electrochemical characterization of pulsed layer deposited hydroxyapatite-zirconia layers on Ti-21Nb-15Ta-6Zr alloy for biomedical application, Applied Surface Science 385 (2016) 368-378. https://doi.org/10.1016/j.apsusc.2016.05.130

[20] I. Gradinaru, I. Stirbu, C.A. Gheorghe, N. Cimpoesu, M. Agop, R. Cimpoesu, C. Popa, Chemical properties of hydroxyapatite deposited through electrophoretic process on different sandblasted samples, Materials Science-Poland. 32 (2014) 578-582. https://doi.org/10.2478/s13536-014-0241-x

Chapter 3

Technology, Methodology and Material Basis used in Experimental Research of Materials for Brake Discs

C. Bejinariu[1], C.D. Florea[1], N. Cimpoesu[1], R. Cimpoesu[1]*

[1]Faculty of Materials Science and Engineering, "Gheorghe Asachi" Technical University of Iasi, Romania

ramona.cimpoesu@tuiasi.ro

Abstract

The research methodology aimed at an experimental planning based on the improvement of the properties of grey cast iron, materials constantly used in brake discs manufacturing. It aimed at characterizing the materials currently used in realization of brake discs, achievement chromium alloy castings, characterizing and controlling the process of growth of thin layers and analyzing and characterizing the new materials. The planning of the experiments included the approach of new analysis techniques that can bring considerable information that participates in obtaining the expected final product. The implementation of the process for making thin layers through thermal spraying of the ceramic powders has the advantage of control of the deposition parameters. The improvement of the properties of the common materials used for the brake discs was achieved by alloying and depositing thin ceramic layers on the contact surface.

Keywords

Methodology, Cast Iron, Ceramic Thin Layers, APS

3.1 Research methodology

Metal brake discs are of particular interest in automotive, railroad or aeronautical applications due primarily to the prices promoted and the technologies already approved and recognized. Their geometric morphology, thickness and the material from which it is made bring great advantages to the sale of this equipment. Based on the thermal conductivity, the special strength and the machinability, the brake discs made of Fe-C alloys will be of special interest in the further development of this necessary field during this period. The analysis of brake discs is not only a gain with immediate applications in the automotive field but an opportunity for many fields such as aeronautics, industrial at any scale. The research presents the approach of the braking systems, from the point of view of the material properties and their influence on automotive industry. As an application, the materials used in the field of stopping systems are experiencing an increasing development and in order to continue in the same manner heat dissipation and friction properties must be developed without large financial investments. The aim is to improve the friction, heat transfer and corrosion resistance properties of the materials for the discs that are part of the braking system. Research follow the purpose to improve the properties of classic materials used in braking discs by advanced processing of metal melting, heat treatment, surface deposition or mechanical processing (laser, CNC) of the braking systems.

The main objective of the work is the development and characterization of metallic materials for cars brake discs based on iron, aluminum, copper or non-metallic ceramic to manufacture special construction types with surface modifications by chemical or mechanical methods.

The research is related to the ability to dissipate the classic materials used for braking systems and its improvement by chemical changes (chemical composition), structural (heat treatment and plastic deformation), surface (deposition of ceramic layers or mechanical processing of surfaces by laser techniques) or by the geometrically constructive modification of the brake disc to increase the heat dissipation capacity obtained from the mechanical one.

The main results presented are: elaboration of a mathematical model that confirms through the linear Tafel and the cyclical curves the electro-corrosion behaviour of the experimental materials; characterization of the classic materials used to make the brake discs; improving the mechanical and chemical characteristics of existing materials by alloying; making thin ceramic layers on cast iron substrate to improve the characteristics of wear and corrosion resistance; analysis of ceramic thin layers for coating metallic materials.

The book has a multidisciplinary character requiring extensive knowledge of materials and mechanical engineering, physics, chemistry and experimental analysis allowing contributions in all areas marked by the presence of applications in the automotive industry, methods of metal deposition and effective uses of finished products that involves superficial ceramic layers. The complex character of the experimental determinations results from the general plan of the research methodology.

Microstructure analyzes were performed by scanning electron microscopy (SEM VegaTescan LMH II electron microscope, secondary electron detector, 2D and 3D structural surface analysis, sizing, light intensity variation and other experimental applications specific to the VegaTescan specialized program), optical microscopy and atomic force microscopy (AFM Nanosurf EasyScan), chemical analysis EDS (dispersive energy spectroscopy) using EDS detector (Bruker, PB-ZAF, automatic or list element and specialized modes of analysis Line, Point or Mapping), and X-ray diffraction (XRD X 'PERT PRO MRD).

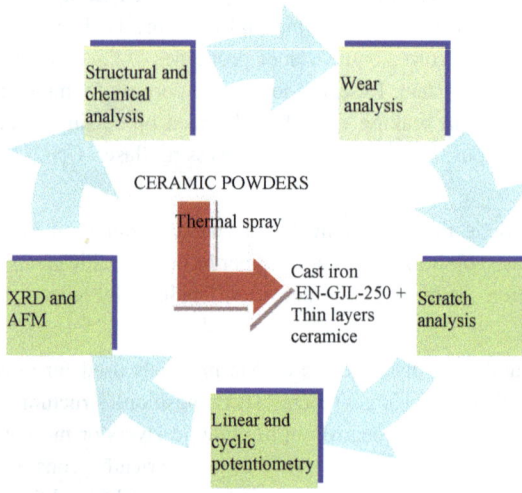

To assess the characteristics of a material proposed for various applications, a group of equipment must simulate, partially or totally the actual conditions in which these materials operate. In this way, we can characterize the proposed materials to increase the time of their use in various braking devices. In this book are presented information about ferrous alloy technology, investigative techniques and equipment used for obtaining and characterization.

3.2   Methodology for research of brake disc materials

The proposed methodology provides the analysis of metallic materials obtained by alloying classical cast irons to increase them resistance to wear, corrosion or change their thermal conductivity.

Also proposed is a second way to modify the characteristics of the basic material (cast iron) by depositing on metallic surface ceramic layers through the Atmospheric Plasma Spraying technique, equipment owned by "Gheorghe Asachi" Technical University of Iasi, the Faculty of Mechanics. The experiments provide the implementation of an experimental architecture for the analysis of the results obtained on the proposed materials, Fig. 3.1, so that after the experimental characterizations, we can select the best material or modify their properties again, according to the results.

In this sense, we propose a chromium alloy of the standard cast iron, currently used on most existing brake discs. After obtaining the alloy and subjecting it to wear systems, the results of corrosion and wear resistance tests, microhardness, profilometry or microstructural and chemical characterization will promote or not this alloy/metal-ceramic system. In the same sense in the second case, that of using superficial layers, we follow the same reasoning and if necessary we will change the deposition parameters such as time, distance, layer thickness when resuming the tests.

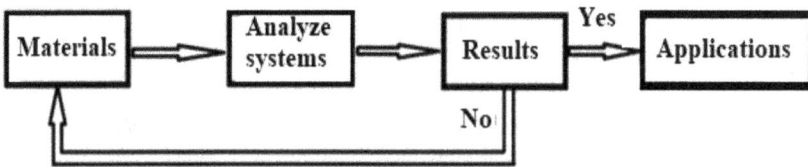

Figure 3.1 The experimental architecture proposed for the investigation of materials for brake discs.

The second variant also has the advantage of a possible reuse of the initial brake disc after it has worn in the contact area and has been re-loaded with ceramic material. For the analysis of the two categories of materials, obtained by alloying and, respectively, by deposition of surface ceramic layers, presented in Fig. 3.2, we propose several investigation techniques that will be described below.

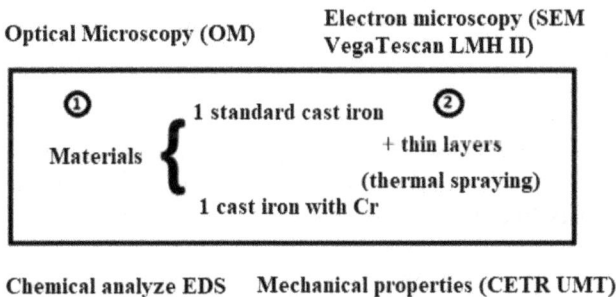

Figure 3.2. Materials proposed in research and analysis techniques.

The microstructures of the materials obtained after alloying and deposition of surface layers by Atmospheric Plasma Spraying (APS), will be investigated through scanning electron microscopy (SEM), also the condition of the surface after the wear tests at the microscopic level. Using VegaTescan software, 2D and 3D surface information can be obtained, a function similar to the results from the atomic force microscope (AFM), even if the roughness of the sample does not allow analysis by atomic force microscopy.

3.3   Design and manufacture of friction alloys used in experimental research

To improve the properties of the alloys used in the construction of brake discs, it was decided to make ferrous materials with the addition of chromium by casting and to deposit ceramic layers on a metal substrate in the contact area of the braking system.

3.3.1   Obtaining cast iron alloys with chrome

To improve the properties of gray cast iron (e.g., EN-GJL-250) with applications in the field of friction systems we opted to alloy these materials with the chrome element. The choice was made primarily due to the recognized anti-corrosion properties introduced by the chromium dioxide that forms on the surface in contact with the surrounding atmosphere. Chromium, from a quantitative point of view, varies in cast iron from the quality of accompanying element (below 0.3%) to the quality of alloying element (over 0.3% - [1] and indicates cast iron with a maximum of 36% Cr which is applied in industry. Experimentally we analyzed materials with 45% Cr. Chromium is an anti-graphitizing element, on a scale, in ascending order of the anti-graphitizing (whitening) effect, the order being the following: W, Mn, Mo, Sn, Cr, V, B, etc. As an anti-graphitizing element, chromium increases the number of eutectic cells and the proportion of perlite, in the case of grey or white cast iron.

At the eutectoid temperature (in the Fe-C binary system) or in the eutectoid temperature range (in the case of industrial cast iron), chromium slows down the diffusion processes and lowers the austenite transformation temperature. Chromium is an α-gen element that favours the development of the occupied domain of p ferrite-in equilibrium diagrams. The perlitizing effect of chromium, from the eutectoid transformation, results from the comparison with other chemical elements, using relative coefficients of influence that have the following values: 1,0 (Cu); 0.2 (Ni); 0.5 (Mn). In principle, chromium has significantly different influences depending on whether the cast iron is weak, medium or high alloy. The chromium content depends on the destination of the cast iron, being 11 - 30% in the case of wear-resistant cast iron, 15 - 25% and 29 - 35% in the case of high temperature-resistant cast iron and 20 - 35% in the case of corrosion-resistant cast iron [2]).

Usually, molybdenum is also found in these castings in a proportion of up to 30%. If the chromium content is less than 9.5%, the structure shows $(Fe,Cr)_3C$ carbide, with orthorhombic network, as part of the eutectic colonies. If the proportion of chromium in cast iron is 9.5 - 13%, in the structure, along with carbide $(Fe, Cr)_3C$, carbide $(Fe, Cr)_7C_3$ also appears, the latter influencing cylindrical and conical eutectic colonies. Carbide $(Fe,Cr)_7C_3$ has a trigonal network and develops toward the longitudinal axis of the cylindrical or conical column of the eutectic one. Structural aspects of the cast iron materials are given in Fig. 3.3 [3].

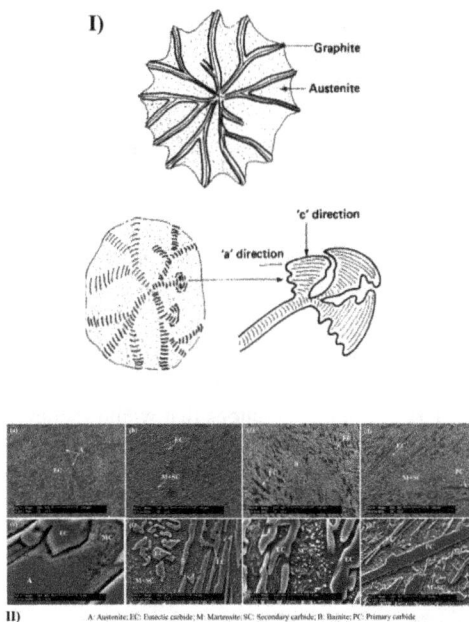

Figure 3.3 Structural aspects I) morphology of the growth of a flake graphite eutectic cell; with the case for a compacted Fe presenting a graphite film evolution along the C axe and also the increase of a round shape of compacted graphite and II) structural aspects of high-Cr cast iron. (a) (e) lower C-Si in as-cast and (b) (f) hardened; (c) (g) higher C-Si in as-cast and (d) (h) hardened [3].

The core of a cylindrical or conical colony - the crystallization germs - is the elongated carbide $(Fe, Cr)_7C_3$. The cylindrical colonies of austenite (austenite envelops the carbide branches) and carbide (Fe, Cr)$_7$C$_3$ grow in the form of bundles. Between the bundles are formed the eutectic based on eutectic cementite -Fe$_3$C- and honeycomb eutectic austenite. If the proportion of chromium is higher than 13%, the cementite disappears from the structure, therefore the eutectic colonies in the form of parallelepiped blocks disappear (honeycomb structure), the eutectic being formed only by cylindrical colonies of austenite and carbide (Fe, Cr)$_7$C$_3$.

In the equilibrium diagram, for hypoeutectic cast irons, there is liquid and austenite in the area between the liquid and solidus curves. For hypereutectic cast iron, in the area between the liquid and solidus curves, there is liquid and carbide (Fe, Cr)$_7$C$_3$, (primary carbide is in the form of hexagons). The metallographic structure of highly chromium-alloy cast iron, in the raw cast state, consists of eutectic carbides in a mass of chromium-alloyed ferrite (for cast iron with a chromium content of less than 13%, there are also perlite phases formed from ferrite and carbides (Fe, Cr)$_3$C). The metal load consists of high-purity steel waste, own waste, ferrochrome and ferromolybdenum or molybdenum oxide.

Chromium alloying is done only in the oven. Chromium losses during processing are a maximum of 5%. A characteristic of chromium alloys is the alloying with nitrogen, an alloy that is made by using ferrochrome master alloy containing nitrogen and introduction into the metal bath in the furnace of the sodium Ferro cyanide; introduction into the casting pot of the urea, sodium nitrate, nitrate potassium, calcium cyanamide, hexamethylenetetramine, ammonia, a mixture of ammonium chloride and sodium nitrite, etc. [4]. In practice it is preferred to use some compositions as close as possible to the eutectic ones due to large liquid contractions that would require the use of large weights (shrinkage to solidification is close to that of steels) [5].

Cast alloys with 24 - 30% chromium are recommended to be inoculated with 0.05 - 0.1% Al, in which case the sulphides are spheroidized (most characteristics are improved) or with 0.2% ferrocerium (the Sulphur content is reduced by 20%, the structure is finished and the non-metallic inclusions are spheroidized). It is recommended that the casting temperature to be at least 1400 °C due to the compact oxide film located on the surface of the liquid cast iron, that causes the formation of films on the surface of the castings, but, in principle, as small as it is possible. High-chromium alloys can be cast in raw, dry and permanent forms [6].

3.3.2 Obtaining ceramic layers

To improve the properties of materials currently used in the brake discs for the automotive industry, it was proposed to deposit ceramic surface layers by atmospheric plasma spraying. The equipment used to obtain the coating is a SULZER METCO 9MCE, system that can make industrial coatings of large or small metal surfaces, depending on the application. Table 3.1 shows the technological parameters used for the coating process. We can see the deposition equipment (spray part) in the Fig. 3.4a) and Fig. 3.4b) by the robotic arm and the rotating sample holder. For the microstructural, chemical and electrochemical experiments, samples with different dimensions were made for depositing ceramic layers using the mentioned method. For determinations of microstructure and chemical composition, cylindrical samples (diameter of 10 mm and length of 3 mm) were realized, Fig. 3.4c).

Table 3.1. Technological parameters used in the submission process of ceramic powders

| Powder | Tun | Ar | | H₂ | | Electric parameters | | Powder supply 9MP | | | Spray distance (Inches) |
		Pressure (psig)	Gas flow (SCFH)	Pressure (psig)	Gas flow (SCFH)	DC (A)	DC (V)	Gas flow transport (SCFH)	Air Pressure (psig)		Speed (lb/h)
Al₂O₃	9MB	75	111	50	10	500	60-70	13.5	20	5.5	3.5
Al, Zr, Y Oxides	9MB	75	110	50	10	500	75	14	20	6	3

psig: kilogram per square inch; SCFH: standard cubic feet per hour; DC: direct current; A: amps; V: volts; lb / h: pound / hour.

For the samples analyzed in terms of resistance to electrochemical corrosion, they were insulated in Teflon, with a diameter of 15 mm, thus allowing an active surface of 314 mm². For the tests of hardness and behaviour at scratching and wear, experimental ceramic layers were made on samples with the following dimensions 50x10x5 mm, cut with wire from the main deposited sample, Fig.3.4 c). The plasma collection system is located outside the depressurized deposition chamber.

The experiments were performed for two sets of samples, with two layers of ceramic material (approximately 30 μm thick/15 μm per layer) and with four layers of material (approximately 60 μm thick) [7-9]. Before the deposition process, the surface of the samples was processed by sandblasting to improve the adhesion of the ceramic layer to the metal substrate [10-14]. For the deposition process, the equipment can cover large areas up to 4 m² in a short time, representing a suitable solution for industrial applications. A rotating support table helps the deposition process to be faster for different experimental substrates (alloys, different shapes or sizes).

The experiments were performed in accordance with the occupational health and safety laws and regulations in order to eliminate all the risks and dangers which can affect the human resource during the experiment procedures [15-17].

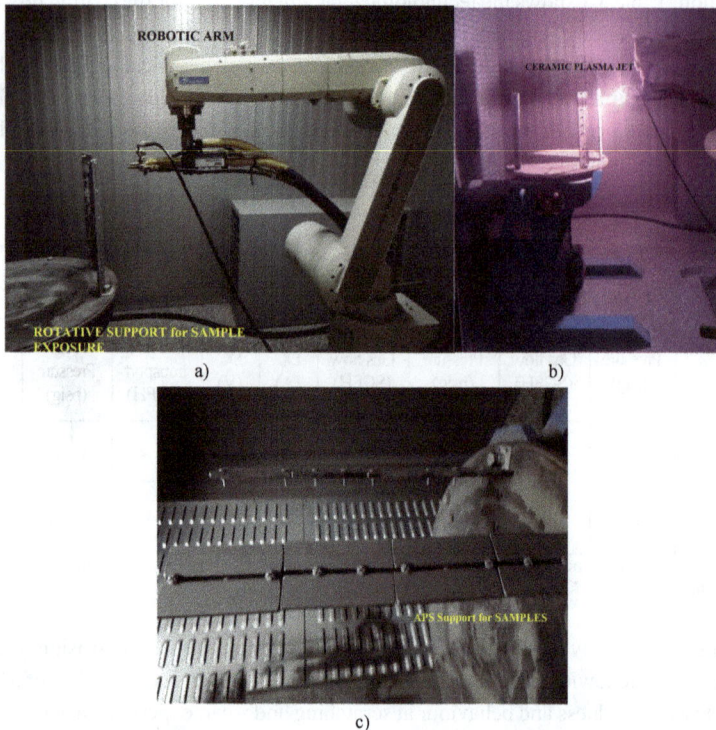

Figure 3.4. Thermal spray deposition equipment:
a) robotic arm; b) submission process; c) support for experimental samples.

References

[1]   H. Ding, S. Liu, H. Zhang, J. Guo, Improving impact toughness of a high
       chromium cast iron regarding joint additive of nitrogen and titanium, Mater.
       Design. 90 (2016) 958-968. https://doi.org/10.1016/j.matdes.2015.11.055

[2]   T. Sun, R.-Bo Song, X. Wang, P. Deng, C.-J Wu, Abrasive Wear Behavior and
       Mechanism of High Chromium Cast Iron. J. Iron Steel Res. Int. 22 (2015) 84-90.
       https://doi.org/10.1016/S1006-706X(15)60014-0

[3]   L. Chen, J.Zhou, V. Bushlya, J.-E. Stahl, Influences of micro mechanical property
       and microstructure on performance of machining high chromium white cast iron

with cBN tools, Procedia CIRP 31 (2015) 172 – 178.
https://doi.org/10.1016/j.procir.2015.03.092

[4]   M. Pokusová, Abrasion Resistance of as-Cast High-Chromium Cast Iron,
Scientific Proc. 2014, Faculty of Mechanical Engineering, SUT in Bratislava. 22,
2014, 74-79. https://doi.org/10.2478/stu-2014-0013

[5]   T.T. Matsuo, C. S. Kiminami, W. J. Botta Fo, C. Bolfarini, Sliding wear of spray-
formed high-chromium white cast iron alloys, Wear. 259 (2005) 445-452.
https://doi.org/10.1016/j.wear.2005.01.021

[6]   G. Xie, H. Sheng, J. Han, J. Liu, Fabrication of high chromium cast iron/low
carbon steel composite material by cast and hot rolling process, Mater Design. 31
(2010) 3062-3066. https://doi.org/10.1016/j.matdes.2010.01.014

[7]   C.D. Florea, I. Carcea, R. Cimpoesu, S.L. Toma, I.G. Sandu, C. Bejinariu,
Experimental Analysis of Resistance to Electrocorosion of a High Chromium Cast
Iron with Applications in the Vehicle Industry. Rev. Chim.-Bucharest. 68 (2017)
2397-2401. https://doi.org/10.37358/RC.17.10.5893

[8]   A. Arghirescu, C. Baciu, N. Cimpoeşu, Experimental Results on Micrometric
Profile of Substrate and Thickness and Roughness of Deposited Layers Through
Thermal Spraying, Advanced Materials Research. 814 (2013) 49-53.
https://doi.org/10.4028/www.scientific.net/AMR.814.49

[9]   M. H. Tierean, L. Baltes, M. Luca, A. Banea, Measurements of dynamic Young
modulus of AlSi10Mg alloy cast in vibrating field, Journal of Optoelectronics and
Advanced Materials 17 (2015) 1868-1873

[10] A.Pascu, E.M. Stanciu, I. Voiculescu, M.H. Tierean, I.C. Roata, J.L. Ocana,
Chemical and Mechanical Characterization of AISI 304 and AISI 1010 Laser
Welding, Materials and Manufacturing Processes 31 (2016) 311-318.
https://doi.org/10.1080/10426914.2015.1025970

[11] A.Pascu, E.M. Stanciu, C. Croitoru, I.C. Roata, J.M. Rosca, N. Cimpoesu, M.H.
Tierean, C. Bogatu, Pulsed Laser Cladding of NiCrBSiFeC Hardcoatings Using
Single-Walled Carbon Nanotube Additives, Journal of nanomaterials, 2019 (2019)
2401295. https://doi.org/10.1155/2019/2401295

[12] R. Vidu, A.M. Predescu, E. Matei, A. Berbecaru, C. Pantilimon, C. Dragan, C.
Predescu, Template-Assisted Co-Ni Nanowire Arrays, Nanomaterials 9 (2019)
1446. https://doi.org/10.3390/nano9101446

[13]   C. Predescu, A.C. Berbecaru, G. Coman, M.G. Sohaciu, A.M. Predescu, E. Matei, R.E. Dumitrescu, S. Ciuca, I.A. Gherghescu, Corrosion Resistance Evaluation of Some Stainless Steels Used in Manufacture of Hydraulic Turbine Runner Blades, Revista de Chimie 70 (2019) 2491-2496. https://doi.org/10.37358/RC.19.7.7367

[14]   V. Nedeff, C. Bejenariu, G. Lazar, M. Agop, 2013. Generalized lift force for complex fluid. Powder Technol. 235 (2012) 685–695. https://doi.org/10.1016/j.powtec.2012.11.027

[15]   C. Bejinariu, D.C. Darabont, E.R. Baciu, I. Ionita, M.A.B. Sava, C. Baciu, Considerations on the Method for Self Assessment of Safety at Work. Environ. Eng. Manag. J. 16 (2017) 1395–1400. https://doi.org/10.30638/eemj.2017.151

[16]   D.-C. Darabont, R.I. Moraru, A.E. Antonov, C. Bejinariu, Managing new and emerging risks in the context of ISO 45001 standard. Qual.-Access Success, 18, (2017) 11–14.

[17]   D.C. Darabont, A.E. Antonov, C. Bejinariu, Key elements on implementing an occupational health and safety management system using ISO 45001 standard. In 8th International Conference on Manufacturing Science and Education (MSE 2017) - Trends in New Industrial Revolution, Bondrea, I., Simion, C., Inta, M., Eds., E D P Sciences: Cedex A, 2017, Vol. 121, p. UNSP 11007. https://doi.org/10.1051/matecconf/201712111007

Chapter 4

Methods and Equipment's for Disks Brake Investigations

N. Cimpoesu[1], C.D. Florea[1], R. Cimpoesu[1], C. Bejinariu[1]*

[1]Faculty of Materials Science and Engineering, "Gheorghe Asachi" Technical University of Iasi, Romania

costica.bejinariu@tuiasi.ro

Abstract

The mechanical properties of materials are an important aspect in choosing certain alloys for applications. Working methodology and equipment used for experimental determination of the friction coefficient and for the disc surface profiles experimental alloys, is presented in this chapter. Also, the working methodology and equipment used to determine the wear resistance of disks brake is presented. The analysis of corrosion resistance was performed by linear and cyclic potentiometry to quickly determine the behaviour of experimental materials. The properties of disks brake materials are due to their chemical composition and microstructure. There are several research methods for determining the composition, most of them spectrometric presented in this chapter.

Keywords

Disks Brake, Wear, SEM, AFM, EDS, XRD, Tafel

4.1 Working methodology and equipment used for the experimental determination of the friction coefficient of alloys

Some of the properties of metallic materials are determined by the indentation test. The indentation operation is classified according to the dimensions of the penetrators, the pressing forces and the depth of penetration into the micro indentations. A micro indentation can be arbitrarily defined as an indentation that has a diagonal of less than 100μm, noting that increased interest is given to indentations with diagonals below 10μm. The force that will be applied to an indenter to produce indentations of such dimensions is important to design and operate a hardness tester but not necessarily to the mechanism used in the indentation process [1].

It is necessary to study the effort made for macro indentations, based mainly on the geometric similarity principle. This principle is fundamental for testing macroindentation and, although it does not guarantee that it will be able to be applied to the smallest indentations, the principle should not be easily abandoned.

Some investigations from the research were done as Brinell macroindentations. Only pyramidal indentations, especially Vickers and Knoop, are used in microindentation tests, but the general principles resulting from Brinell indentation studies can still be applied due to the geometric similarity of indentations [2-4].

A micro-nano tribometer CETR UMT equipment was used for the experiments, Fig. 4.1a). This is an equipment in modular construction with servo control for the main parameters. It has the possibility to be equipped with: disk pin test systems at micro and nano scale; micro and nano scale wear monitoring systems; micro and nanoindentation, micro and nano scratch systems; a monitoring system during friction tests or wear of lubricant layers or other deposited layers and acoustic emission monitoring system (AE: the frequency range of the sensor is between 0.2 MHz - 5 MHz and the amplification degree of the AE signal is maximum 60 dB) of the friction and wear processes. The research used micro-scale analysis modules for the base materials and the deposited thin layers.

The analysis system will be used to determine the following properties:
  - frictional forces and the static and dynamic micro-scale friction coefficients in the rotational movement for various combinations of materials;
  - the study of stick-slip processes at the micro scale;
  - the study of micro-scale adhesion forces;
  - the study of wear processes at the micro scale;
  - the study of the resistance to micro scratching of the ceramic surface layers and to the determinations of hardness and modulus of elasticity by micro indentation.

The values for the measurable pressing and frictional forces are between 0.1mN and 20N, with a resolution between 1 μN and 1 mN, depending on the measuring range of the force sensors. It is required to cover the following force fields: 0.1 ÷ 10 mN; 5 ÷ 500 mN; 0.2 ÷ 20 N. The apparatus is equipped with a system for displacing the specimen and for controlling the force in the vertical direction (Z direction) for the loading force of the specimen (pin / ball) - with the possibility of programming the loading force (continuous, stepped, fixed) with the following characteristics: maximum stroke 150 mm; travel accuracy of 0.5 μm; travel rate: 0.002 ÷ 10 mm/s; monitoring the depth of the wear trace with an accuracy of up to 5 microns.

Analyzes of the indentation mechanism on blunt pyramid materials were based on a slip-line solution initially developed by Hill, Lee, and Tuper [5]. This process is a two-dimension treatment of an ideal rigid-plastic material. This material is perfectly rigid to the limit of flow, but with very high plastic deformation capacity. The application of this deformation operation, but also several subsequent solutions were reviewed by Shaw [1]. Fig. 4.1a) shows the CETR UMT equipment for determining mechanical properties and the scratching angle applied on the surface, Fig. 4.1 b).

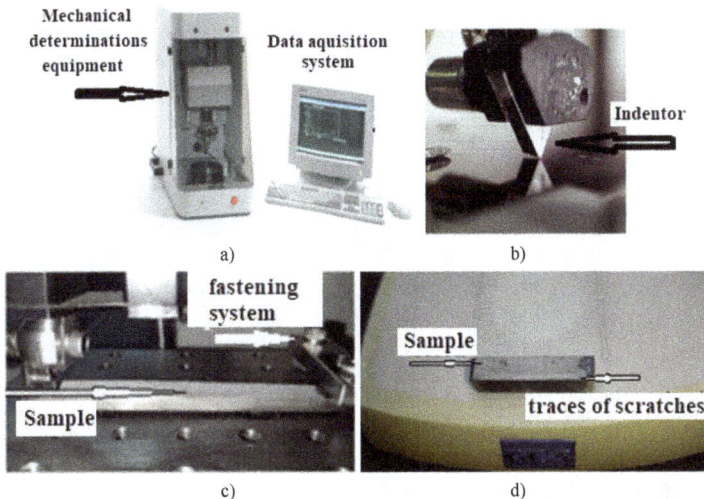

Figure 4.1. CETR UMT equipment for determining mechanical properties:
a) experimental device;
b) the scratch angle was slightly changed on the left side by the friction effect;
c) the experimental sample caught in the fastening system;
d) the image of the sample after the scratch test.

The solutions of the scratch behaviour are based on the fact that the indentation cuts the samples on the plane of the sample surface. As a result, two new surfaces are created, which rotate around the point of contact during indentation. The original material located at the starting point is thus moved to a new point, and the material in the volume of the scratch mark is plastically deformed on the side with upward movements. Relative movement is required between the indentation surface and the sample, therefore friction should have a significant effect, which can be considered by the calculation variant. The effort exerted on the sample material increases the coefficient of friction.

Samuels performed a detailed quantitative analysis of indentations made of cones or pyramids with various angles, investigating materials whose mechanical properties are close to those of an ideal elasto-plastic solid - a solid that is elastic to a point then it deforms plastically with the same difficulty [6]. The slip domain theory can easily be considered applicable to such a material. Samuels established that these indentation characteristics are conform with the model when the indenter angle was less than 60 degrees [6,7]. Indentations were made with tips that had larger angles, however, they did not conform with theory at a hundred percentage fit. The evaluations made were correct because:

    • a cutting (piercing) mechanism needs the elasto-plastic limits do not exceed the peak of the indentation. In practice, the deformation extends over a considerable depth beyond the tip of the indentation;

    • a cutting (piercing) mechanism proposes that the movement of the points on the surface of the sample has a large component parallel to the surface. In practice, the displacement is relatively small;

    • if the cutting occurred, the height of the raised material adjacent to the indentation would be approximately one -third of the indentation depth. In practice, it is only a fraction of it, never more than half of what is needed.

The discrepancies between predictions and observations increase as the wedge angle exceeds 120° and becomes marked at a wedge angle of 140°.

4.2 Working methodology and equipment used for the experimental determination of disc surface profiles of experimental alloys

The surface condition, in the case of two materials that come into contact, is important for the coefficient of friction that occurs between them due to the specific behaviour at the micrometric scale of each metallic or non-metallic material. The surface topography can be evaluated by microscopic methods (at the roughness scale) or nanoscopic methods (at the atomic and molecular scale). Microscopic methods, especially mechanical and optical methods, are sufficient to study the surfaces resulting from processing. Measurement techniques can be divided into two categories:

- direct contact measurement when the surface is touched by a probe;
- contactless measurement, when the surface is scanned by an optical device.

The first direct contact measuring instrument was developed by Abbott and Firestone in 1933. In 1939 Rank Taylor Hobson of Leicester England introduced the first commercial instrument called Talysurf. Today, direct contact measuring instruments and electronic amplification are the most common. The technique with direct contact measurement is recommended by ISO 9001 and is used as reference. In 1983 it was developed a contactless optical profilometer that uses the principle of interferometry of two optical beams. It is widely used in the electronics and optical industry to measure smooth surfaces.

The measurement techniques were divided into six categories according to the physical principle of operation: a mechanical probe method, optical methods, microscope sample - scanning methods (SPM), fluid methods, electrical methods and electron microscopy methods. The construction of the analogue position transducer (with variable inductance) used in many models of measuring instruments, Fig. 4.2 a), has the principle diagram represented in Fig.4.2 b).

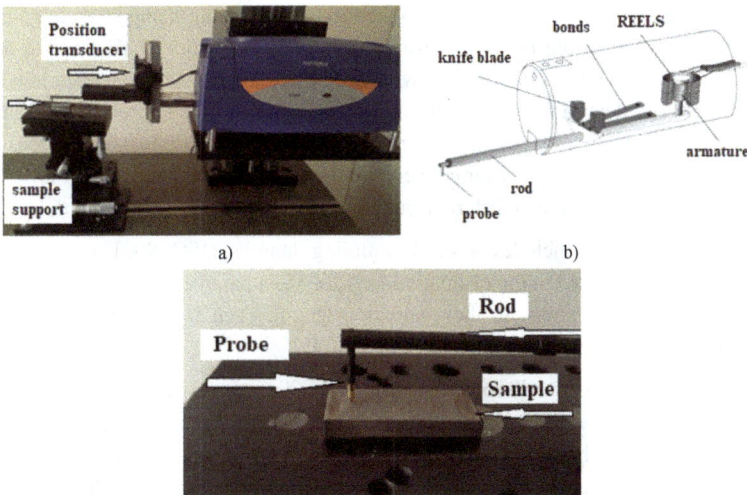

Figure 4.2 Taylor Hobson Profilometer: a) equipment; b) position transducer scheme; c) the experimental sample with a ceramic layer.

The tip is supported at one end of a rod that pivots on the edges of the knife, which are in the form of a prism. The farther end supports an armature that moves between two coils, changing the relative inductance. The coils are connected to a circuit in the AC bridge, so that when the armature is in the center the bridge is balanced and gives no output signal. The movement of the reinforcement unbalances the bridge, which produces an output signal proportional to the movement, the relative phase of the signal depending on the direction of movement. The signal is amplified and compared to that of an oscillator to determine in which direction it moved from the center position (zero).

The tip is kept in contact with the surface and edges of the knife blades, Fig. 4.2b) with the help of a weak spring acting on the bar. The links prevent the bar from moving horizontally, resulting in the free movement of the probe only in the vertical direction. The electrical output signal represents the movement of the probe and the transducer body relative to a surface level. Therefore, the transducer must cross along a line parallel to the surface so that the electrical output signal is a real representation of the profile being followed. The movement of the armature is converted into an electrical signal in the transducer coil. This signal is processed and amplified in analogue electronics before being considered digital (digitized). The bandwidth of the electronic part must be sufficient to adapt to all the characteristics of the measuring surface, with good linearity and signal/noise ratio without distortion. Several determinations can be made on the surface of the experimental samples. Sample length (L_c) is the length on the surface over which a single parameter assessment is made (e.g., Ra) and is chosen, for convenience, the same as the cut-off length [8]. Some instruments require choosing a different sampling value from the sample length.

The evaluation length includes several sampling lengths (ISO 9001 recommends 5 sampling lengths included in the evaluation length). Most parameters are calculated as the mean value over all samples in the evaluation length, even if in some cases it is the maximum or minimum value in any sample length [9].

The measuring length is the distance over which the data are processed after filtering the signal, a certain volume of data being removed, in this way the measuring length remains the evaluation line of the test. The dimension of the piece represents the distance that the probe crosses the surface during the experiment. The cross member is longer than the measuring length as long as necessary to allow a small overtaking due to mechanical acceleration and deceleration. With Taylor Hobson's Form Talysurf, these distances are 0.3 mm at the beginning and 0.1 mm at the end. The measurement of an area can be affected if the wavelength is not limited correctly. The upper limit is set by the length of the sample. The short wavelength limit (high frequency) is imposed by the design of the measuring instrument, the dimensions of the probe tip, the electrical or mechanical response of the measuring system, or the sampling rate at which the profile data is digitized before being processed by computer.

To more accurately describe the response of an instrument, the bandwidth is used between the sampling length (L_c) and the shortest detectable wavelength (L_s). This is the bandwidth instrument and normally it is expressed as the ratio between the two limit wavelengths, L_c/L_s. The ISO 9001 recommendation is that a bandwidth of 300: 1 should be used. For example, when a sampling length of 0.08 mm is used, values of sampling lengths of less than 0.0025 mm are used for bandwidths of 300:1. Although the parameters Ra and σ are the most used in the technical specifications of the machine parts, all the mentioned parameters provide indications regarding only the relative deviations on the profile height, without providing any indications regarding the slope, shape or frequency. Surfaces that have completely different profiles in shape or frequency may provide the same values for the R_a or σ average parameters.

The value of the skewness parameter depends on how the solid material is distributed within the profile, relative to the mean line: if it is above the mean line, then the value of the skewness parameter is negative, and if the solid material is below the mean line, then the value of the parameter skewness is positive, Fig. 4.3.

$$Sk = \frac{1}{\sigma^3 \cdot N} \sum_{i=1}^{N} (z_i - m)^3$$

(4.1)

where: Sk   -   Skewness parameter;
        σ   -   standard deviation;
        N   -   number of samples;
        i   -   1...N;
        z   -   roughness Height;
        m   -   distance between the reference line and the midline.

A symmetrical distribution of heights determines a number of peaks equal to that of the distributions. Thus, the skewness parameter can be used to differentiate surfaces that have the same value for the arithmetic mean height although they are characterized by a much different shapes of profiles. Roughness at which the tips have been removed or that have deep scratches, leads to negative values of the skewness parameter. Profiles characterized by high peaks or at which the depressions are gentle have positive values of the skewness parameter.

The kurtosis parameter also appreciates the shape of the profile.

$$K = \frac{1}{\sigma^4 \cdot N} \sum_{i=1}^{N} (z_i - m)^4$$

(4.2)

where: K   is   the kurtosis parameter;
        σ   -   standard deviation;
        N   -   number of samples;
        i   -   1...N;
        z   -   height of roughness;
        m   -   the distance between the reference line and the midline.

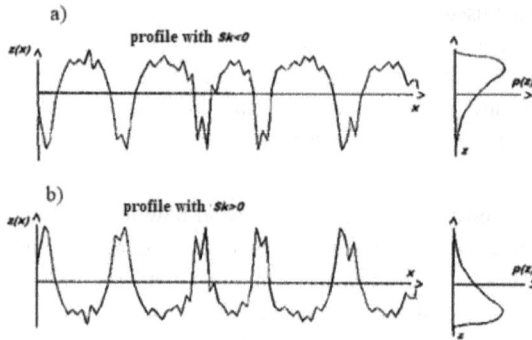

Figure 4.3. Distribution of solid material within the profile: a) parameter Sk<0; b) parameter Sk>0.

Thus, if within the limits of the reference length the profile has relatively few high peaks and depressions, Fig.4.4, the result is a value of K<3, platykurtic profile, and if on the contrary the profile has many high peaks and deep valleys, the result is k>3, leptokurtic profile.

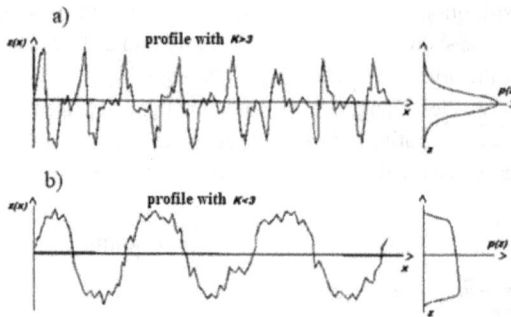

Figure 4.4 Distribution of solid material within the profile: a) parameter K>3; b) parameter K<3.

As we have seen before, the skewness parameter is useful in evaluating the asymmetry of the roughness height distribution in relation to the mean line. The symmetric distributions and the Gaussian distribution have the value zero for the skewness parameter, while the asymmetric distribution in relation to the mean line of the roughness height also determines an asymmetric shape of the probability density function [10].

4.3 Working methodology and equipment used to determine the wear resistance of experimental alloys

To analyze the wear behaviour, metallic and metallic materials with ceramic coatings were subjected to friction wear tests in linear contact couplings, on the Amsler type device, from the Faculty of Mechanics, Technical University Gheorghe Asachi from Iasi, Romania, presented in Fig. 4.5. Parallelepiped samples (50x20x20 mm) were taken from the experimental alloys, which were mounted in the shoe holder device, the friction being made in linear contact by rolling a steel disc with a hardness of 71.5 HRC. The tests were performed at the disk rotation frequency of 20 Hz, at a system load of 12.5 kg for 15 minutes.

Data acquisition for friction moment estimation is done with transducers and Vishay P3 equipment [11,12]. LabView software was used to register and analyze the results and process the statistical data. Using this equipment, the statistical parameters showed that data acquisition is appropriate with insignificant variation of the data (within the limit of 0.02) and a standard deviation of 0.15 with a value at least twice lower than that presented by experimental data filter. The block diagram made in LabView is given in Fig. 4.6.

The coefficient of friction was calculated with the relation (4.3):

$$\mu = \frac{M_t.100}{G.\cos(\alpha).R},\qquad\qquad(4.3)$$

where: M_t - is the friction moment measured during contact in [Kg.cm]

G - weight force, G= 20 [N];

α - angle between the normal axis at the contact surface and the axis of the loading force, $\alpha=25^0$;

R - radius of the friction cylinder, R=24.9 [mm].

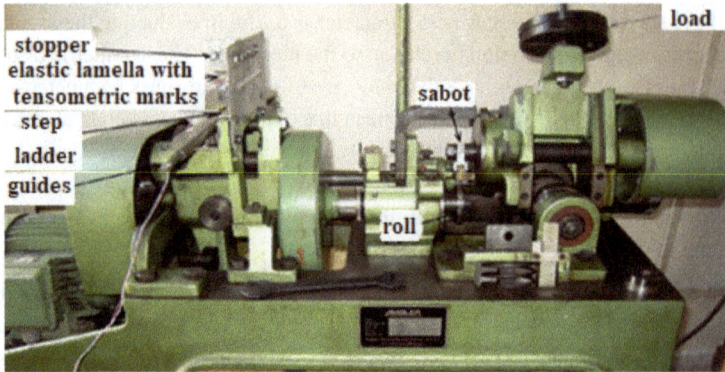

Figure 4.5 Mechanical testing equipment: a) AMSLER assembly; b) detail of the wear area.

4.4   Working methodology and equipment used for experimental determination of corrosion resistance of experimental alloys

The analysis of corrosion resistance was performed by linear and cyclic potentiometry to quickly determine the behaviour of experimental materials. Potential determinations in an open circuit and dynamic polarization were performed using Volta Lab 21 - equipment (Radiometer, Copenhagen). The calculation of the corrosion rate of a metallic material introduced into an electrolyte involves the determination of the instantaneous current density. The value of the instantaneous current is determined by the polarized resistance method. The resistance method helps characterize the corrosion current involved in the potential corrosion of a metal (alloy). The corrosion current is, in this case, the current that forms at the interface between a metal (alloy) and an electrolyte solution representing the environment, once the material is immersed in the solution.

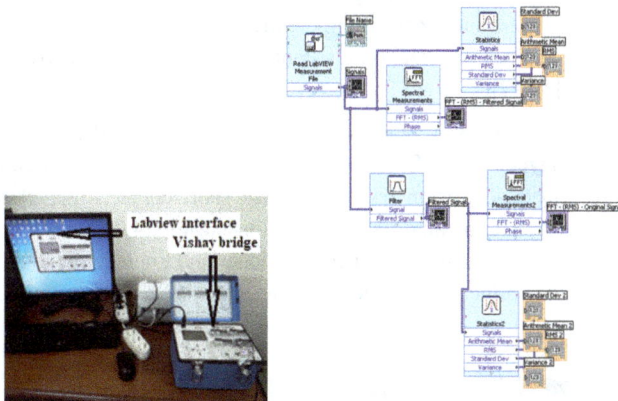

Figure 4.6. Signal acquisition equipment: a) Vishay P3 bridge; b) the block diagram used in Labview for data acquisition and processing.

The average power lost in the friction process, P_f [W], can be estimated using the relation (4.4):

$$P_f = \frac{\pi.G.\cos(\alpha).R}{30}.\mu.N,$$ (4.4)

where: N - is the speed of the friction roller measured using SKF type X-ray equipment, [rpm].

The acquisition and interpretation of data acquired from the work cell with the potentiostat equipment is based on the use of the VoltaMaster 4 software [13]. Experimentally, a glass cell with three electrodes was used. The cell was maintained on a magnetic stirring system to prevent the formation of gas bubbles on the surface of the alloy. Experimental metal samples were introduced into the Teflon supports insulating an active surface of known size. These constituted the working electrode during the experiment. The auxiliary electrode, made of platinum, and the reference electrode, made of saturated calomel, are positioned together with the working electrode at approximately equal distances in the cell with the electrolyte. Experimental determination was performed at room temperature and the electrolyte solution was aerated naturally [14,15].

Linear polarization graphs were made with a potential speed of 1 mV/s for a range of ±150 mV and using the open potential circuit. To determine the cyclic polarization variations, the data were taken at a speed of 10 mV/s in the potential range -750... + 1450 mV. Using the data collection and interpretation program, VoltaMaster, the following electrochemical characteristics were determined: corrosion potential at zero corrosion current E_0 (I = 0), Tafel constants (b_a and b_c), polarization resistance R_p, J_{corr} corrosion current density and V_{corr} corrosion rate [16,17].

Electro-chemical experiments were also used to detect pores and micro-cracks in the deposited ceramic layers. The effects of these surface or layer imperfections on the behaviour of the electro-corrosion resistance for a short contact interval between the sample and the electrolyte medium were also evaluated. Fig. 4.7 shows the potentiostat equipment provided by the Faculty of Materials Science and Engineering in Iaşi, Romania, which was used to determine the corrosion resistance [18,19].

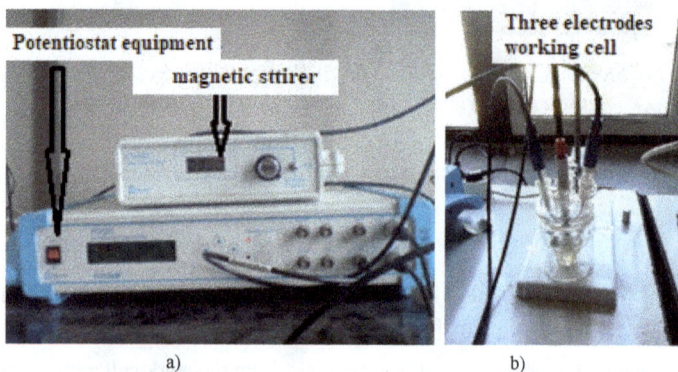

Figure 4.7 Potentiostat device used for the analysis of electro-corrosion behavior: a) equipment; b) cell with 3 electrodes.

The working electrode, allows the mounting of the Teflon fixing and insulation system of the experimental metal sample. The area of the sample exposed to the tests performed was 0.25 cm². The experimental results for potentials are made for the saturated calomel electrode. The anodic polarization test was performed in electrolytic acid rain solution.

4.5 Research methodology of the structure and chemical composition of experimental alloys

The microstructure of materials can be determined by microscopy, which depends on the scale being worked on, can be optical, scanning electronics or transmission electronics. This subchapter presents some operating principles and characteristics of scanning electron microscopy (SEM), analysis of the chemical composition by energy-dispersive X-ray spectroscopy (EDS) or phase analyze by X-ray diffraction (XRD). These techniques and equipment were used to evaluate experimental materials and their wear or corrosion behaviour.

4.5.1 Scanning electron microscopy

The scanning electron microscope (SEM, Fig. 4.8) is an equipment with multiple uses, recently, in research but also in industry due to the many applications that currently use micron-sized structures that cannot be analyzed by optical microscopy.

The quality and resolution of SEM images depend on the following major parameters: device performance; selection of image parameters (for example, the operator control); the nature of the specimen [20].

All three aspects simultaneously work and none of them should or cannot be ignored or overexposed. One of the surprising aspects of scanning electron microscopy is the apparent simplicity with which SEM images of three-dimensional objects can be interpreted by any observer unfamiliar with the device. This thing is surprising, given the unusual way the image is formed, which differs greatly from the normal human experience with images formed by light or eyes.

The main components of a typical SEM are: electron cannon; column of electro-magnetic lenses; scanning system; the detector (s); vacuum system and electronic controls. To produce images, the primary electron beam (consisting of a tungsten filament) is concentrated in a small area (90 nm), with which the surface of the specimen is scanned using scanning coils (electromagnetic lenses). Each point of the analyzed sample bombarded by accelerated electrons will emit a signal in the form of electromagnetic radiation. Only selected elements of this radiation, secondary electrons (SE) and/or backscattered electron (BSE), are collected by a detector and the resulting signal is amplified and displayed on a TV screen or a monitor. The resulting image is simple to interpret, at least for topographic imaging of objects at low amplification powers. The electron beam interacts with the specimen to a depth of approximately 1 - 5 µm. The complex interactions of the electrons of the primary beam with the atoms of the sample produce a wide variety of radiation. The need to understand the image formation process for a reliable interpretation of images occurs in special situations and especially in the case of images with high amplification power [21,22].

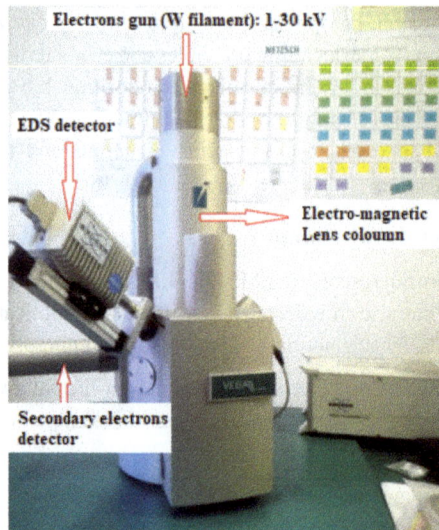

Figure 4.8 Scanning electron microscope VegaTescan LMH II.

Because the SEM microscope is operated under high vacuum, the samples that will be studied must be compatible with a high vacuum ($\sim 10^{-5}$ mbar) without destroying them. This means that liquids and materials containing water and other volatile components cannot be studied directly. Also, samples of the fine powder type must be firmly attached to a substrate of the sample holder so that they do not contaminate the sample chamber of the SEM. Non-conductive materials must be attached to a conductive sample holder and covered with a thin conductive film by spraying or evaporation. Typical coating materials are Au, Pt, Pd, their alloys, and carbon.

4.5.2 Dispersive energy spectrometry

The EDS analysis technique uses the X-ray spectrum emitted by a solid sample bombarded with a primary electron beam to obtain a localized chemical analysis. All chemical elements from atomic number 4 (Be) to 92 (U) can be detected in principle using this technique, although not all devices are equipped for light elements (Z <10). Qualitative analysis involves the identification of lines in the spectrum and is easy to obtain due to the simplicity of the X-ray spectrum. Quantitative analysis (determining the concentrations of the present elements) involves measuring the intensities for each element in the sample.

Figure 4.9 Bruker EDS detector for qualitative and quantitative chemical determinations.

During the research, various EDS analyses were performed to characterize the deposits made, the effects of scratch or wear tests, the effects of corrosion resistance tests of metallic materials or complex ones with ceramic surface layers. The detector used on an SEM system is from Bruker (Silicon Drift), shown in Fig. 4.9.

Analyses can be performed in Automatic or Element List mode on selected surfaces in the sample or specific analysis modes can be used, for example: Point mode (point chemical analysis - on the surface of a 90 nm spot), Line mode (L - distribution of elements identified on a user-selected line) and Mapping mode (M - distribution of chemical elements on the analyzed area) [23].

X-ray intensities are measured by photon counting, and the accuracy obtained is limited by the statistical error of the detector. An accuracy of ± 1% is obtained for the main elements, but the overall analytical accuracy is closer to ± 2% due to other factors, such as uncertainties in the standard compositions and errors in the various corrections to be applied to the raw data. The EDS detector through the operating program also displays the error specific to each chemical element identified according to its percentage in the analyzed sample.

gBecause the electron detector analyses only at a shallow depth in the experimental sample, the specimens will be well polished so that the surface roughness does not affect the results. In principle, samples of any size and shape can be analyzed (within reasonable limits determined by the vacuum enclosure of the equipment). Many samples are not electrically conductive, and a coating must be applied to the conductive surface to provide a pathway for the incident electrons to be transmitted. The commonly used coating material is carbon vaporized under vacuum (with a thickness of ~ 10 nm). It has a minimal influence on X-ray intensities due to its low atomic number and, unlike the gold commonly used for SEM specimens, does not add unwanted peaks to the X-ray spectrum.

The lowest detectable peak for a chemical element can be defined as three times the standard deviation of the background noise. An estimate of the detection limit of the order of magnitude can be obtained as follows: if the counting speed for a pure element is 1000 identifications/s and the ratio between the peak and background noise is 500:1, the noise counting speed background is 2 identifications/s. In 100 seconds, 200 background counters will be accumulated, giving a relative standard deviation of $(200^{1/2}/200)$ or 0.07.

Since the background intensity in this case is equivalent to the peak counting speed for a concentration of 1000 ppm, three standard deviations are equivalent to a concentration of 0.07 x 3 x 1000 = 210 ppm.

Reducing the detection limit requires more metering, which can be achieved by increasing the counting time and / or beam current. In EDS analysis, detection limits are typically approximately 0.1%, although a reduction can be achieved using longer counting times or changing the counting speed higher [24,25]. The values given here for the limits of detection refer to samples such as silicates, for which the atomic number (which determines the continuous intensity) is low. Phases that contain heavy elements offer higher detection limits due to the higher background.

4.5.3 X-ray diffraction

X-ray diffraction can be used to determine the crystal structure of materials and the parameters of characteristic networks. This information can then be used to identify the analyzed material because each metal element in the periodic table has a unique combination of network structure and its parameters at room temperature. When an X-ray beam is directed at a metal crystal, the beam strikes the atoms and produces two types of radiation: white and characteristic X-rays [20,26].

White X-rays include several wavelengths and are of no interest in this experiment. The characteristic X-rays are caused by the ejection of an electron from an inner layer of an atom hit by the incident X-ray. When an electron in the outer layer moves to fill the space created in the inner layer, energy is emitted in the form of a photon of X-rays. Bragg's law is used to determine the crystal parameters based on its characteristic X-ray pattern. Within the experiments performed in the paper, an X Pert device was used, Fig. 4.10.

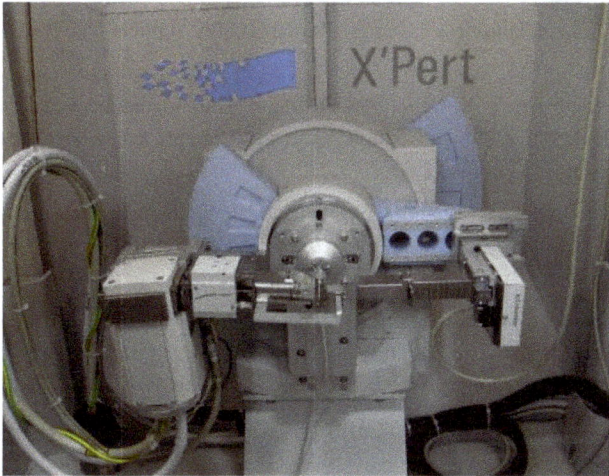

Figure 4.10 X Pert equipment for the analysis of experimental materials.

In this research, the XRD spectra of the EN-GJL-250 cast iron material and of the experimental material covered with ceramic surface layers were analyzed. X-rays that come in contact with a crystal have approximately the wavelength close to the space between the atoms in the crystal lattice. Bragg's law can be derived by considering a cubic crystal lattice consisting of parallel planes of atoms. If we consider that each plane acts as a surface that is hit by the incident X-ray beam, we see that the beam is reflected in some cases and in others not. In the case of reflection, we observe that the beams coming out of the crystal are in phase and act to strengthen each other. This situation happens when the beam of the incident strikes parallel planes at certain angles known as Bragg Angles, θ. In the non-reflecting case, waves leaving the crystal are out of phase and cancel each other out [27,28].

References

[1]   P.J. Blau, Methods and Applications of Microindentation Hardness Testing. In: Vander Voort G.F. (eds) Applied Metallography. Springer, Boston, 1986. https://doi.org/10.1007/978-1-4684-9084-8_9

[2]   H. Chandler, Hardness testing, second edition, ASM International, 1999.

[3]   K. Herrmann, Hardness Testing – Principles and Applications, ASM International, 2011.

[4]   T. Wilantewicz, W.R. Cannon, G.D. Quinn, The Indentation Size Effect (ISE) for Knoop Hardness in Five Ceramic Materials. Ceramic Engineering and Science Proceedings. 27 (2006) 237 – 249. https://doi.org/10.1002/9780470291368.ch20

[5]   J. Grunzweig, I.M. Longman, N.J. Petch, Calculations and measurements on wedge-indentation. J. Mech. Phys. Solids. 2, (1954) 81-86. https://doi.org/10.1016/0022-5096(54)90002-2

[6]   L.E. Samuels, T.O. Mulhearn, An experimental investigation of the deformed zone associated with indentation hardness impressions. J. Mech. Phys. Solids. 5 (1957) 125-134. https://doi.org/10.1016/0022-5096(57)90056-X

[7]   G. Tu, S. Wu, J. Liu, Y. Long, B. Wang, Cutting performance and wear mechanisms of Sialon ceramic cutting tools at high speed dry turning of gray cast iron. Int. J. Refract. Met. H. 54 (2016) 330-334. https://doi.org/10.1016/j.ijrmhm.2015.08.007

[8]   R. C. Bradt, D. Munz, M. Sakai, K. W. White, Fracture Mechanics of Ceramics, Active Materials, Nanoscale Materials, Composites, Glass and Fundamentals, Springer, 2003.

[9]   C. Castillo-Peces, C. Mercado-Idoeta, M. Prado-Roman, C. Castillo-Feito, The influence of motivations and other factors on the results of implementing ISO 9001 standards. European Research on Management and Business Economics. 24, (2018) 33-41. https://doi.org/10.1016/j.iedeen.2017.02.002

[10]   L. Blunt, P. J. Sullivan, The measurement of the pile-up topography of hardness indentations. Tribol. Int. 27 (1994) 69-79. https://doi.org/10.1016/0301-679X(94)90072-8

[11] Z. Zalisz, A.Watts, S. C. Mitchell, A. S. Wronski, Friction and wear of lubricated
M3 Class 2 sintered high speed steel with and without TiC and MnS additives.
Wear. 258 (2005) 701-711. https://doi.org/10.1016/j.wear.2004.09.069

[12] A. Zavos, P. G. Nikolakopoulos, Tribology of new thin compression ring of fired
engine under controlled conditions -A combined experimental and numerical
study. Tribol. Int. 128 (2018) 214-230.
https://doi.org/10.1016/j.triboint.2018.07.034

[13] G. Bolat, D. Mareci, S. Iacoban, N. Cimpoeşu, C. Munteanu, 2013. The estimation
of corrosion behavior of NiTi and NiTiNb alloys using Dynamic Electrochemical
Impedance Spectroscopy. J. Spectrosc. (2013) ID 714920.
https://doi.org/10.1155/2013/714920

[14] D. Mareci, N. Cimpoesu, M. I. Popa, Electrochemical and SEM characterization
of NiTi alloy coated with chitosan by PLD technique. Mater. Corros. 63 (2012)
176-180. https://doi.org/10.1002/maco.201206501

[15] M. Zaharia, S. Stanciu, R. Cimpoeşu, I. Ioniţă, N. Cimpoeşu,. Preliminary results
on effect of H2S on P265GH commercial material for natural gases and petroleum
transportation. Appl. Surf. Sci. 438 (2017) 20-32.
https://doi.org/10.1016/j.apsusc.2017.10.093

[16] S.L. Toma, C. Baciu, C. Bejinariu, D.A. Gheorghiu, C. Munteanu, N. Cimpoeşu,.
Studies on the Corrosion Behavior of Deposits Carried out by Thermal Spraying in
Electric ARC – Thermal Activated. Applied Mechanics and Materials. 657 (2014)
261-265. https://doi.org/10.4028/www.scientific.net/AMM.657.261

[17] N. Forna, N. Cimpoeşu, M.-M. Scutariu, D. Forna, C. Mocanu, Study of the
electro-corrosion resistance of titanium alloys used in implantology. E-Health and
Bioengineering Conference, EHB 2011. art. no. 6150362.

[18] N. Cimpoeşu, L. C. Trincă, G. Dascălu, S. Stanciu, S.O. Gurlui, D. Mareci,.
Electrochemical Characterization of a New Biodegradable FeMnSi Alloy Coated
with Hydroxyapatite-Zirconia by PLD Technique, Journal of Chemistry (2016)
Article ID 9520972. https://doi.org/10.1155/2016/9520972

[19] J. Izquierdo, G.Bolat, N. Cimpoesu, L. C. Trinca, D. Mareci, R.M. Souto,
Electrochemical characterization of pulsed layer deposited hydroxyapatite-zirconia
layers on Ti-21Nb-15Ta-6Zr alloy for biomedical application. Appl. Surf. Sci. 385
(2016) 368-378. https://doi.org/10.1016/j.apsusc.2016.05.130

[20]   C. Munteanu, M. Ştefan, C. Baciu, N. Cimpoeşu, Metode difractometrice şi microscopie optică şi electronică în studiul materialelor, Editura Tehnopress, Iasi, 2008.

[21]   I. Hopulele, N. Cimpoeşu, C. Nejneru, Metode de analiză a materialelor. Microscopie Şi Analiză Termică Editura Tehnopres, 2009.

[22]   L. Reimer, Scanning Electron Microscopy, Physics of Image Formation and Microanalysis, Springer, 1998. https://doi.org/10.1007/978-3-540-38967-5

[23]   J. Heath, Energy Dispersive Spectroscopy, second edition, John Wiley & Sons Ltd 2015.

[24]   R. K. Mishra, A.K. Zachariah, S. Thomas, Chapter 12: Energy-Dispersive X-ray Spectroscopy Techniques for Nanomaterial, Microscopy Methods in Nanomaterials Characterization, 2017, p. 383-405. https://doi.org/10.1016/B978-0-323-46141-2.00012-2

[25]   J.J. Leani, J.I. Robledo, H.J. Sánchez, Energy dispersive inelastic X-ray scattering spectroscopy – A review, Spectrochim. Acta B. 154 (2019) 10-24. https://doi.org/10.1016/j.sab.2019.02.003

[26]   W. Yoshio, M. Eiichiro, S. Kozo, X-Ray Diffraction Crystallography, Introduction, Examples and Solved Problems, 2011, Springer.

[27]   E.N. Maslen, A.G. Fox, M.A. O'Keefe, X-ray Scattering, E. Prince (Ed), International Tables for Crystallography, Vol. C Kluwer Academic, Dordrecht, 2004, p. 554.

[28]   M. E.J. Harper, C. Cabral, P. C. Andricacos, L. Gignac, I. C. Noyan, K. P. Rodbell, C. K. Hu, Mechanism for microstructure evolution in electroplated copper thin films near room temperature, J. Appl. Phys. 86 (1999) 2516. https://doi.org/10.1063/1.371086

Chapter 5

Research and Experimental Contributions on the Characterization of Materials for the Construction of Vehicle Brake Discs

C.D. Florea[1], C. Bejinariu[1], R. Cimpoesu[1], N. Cimpoesu[1]*

[1]Faculty of Materials Science and Engineering, "Gheorghe Asachi" Technical University of Iasi, Romania

nicanor.cimpoesu@tuiasi.ro

Abstract

Analyzing the literature, two directions were approached to improve the properties of metallic materials used in obtaining discs for braking systems. The first is to obtain metallic materials with the addition of chromium and the second is to use thin ceramic layers on the contact surface. In this chapter are presented microstructural (OM, SEM) and chemical composition (EDS) analysis of experimental materials obtained by classical casting and analysis of ceramic thin layers obtained by plasma spraying.

Keywords

OM, SEM, EDS, Brake Discs, Thin Layers

5.1 Experimental research for the microstructural characterisation of metallic materials intended for realization of vehicle brake discs

During the experimental research for the realization of the vehicles braking discs, two categories of metallic materials were approached:
- Fe - C alloys of cast iron type, in cast state;
- clad materials obtained by depositing multiple layers of aluminium oxide (Al_2O_3) on a grey cast iron support.

5.1.1 Physico - chemical characterization of experimental Fe - C alloys

The physico-chemical characterization of the experimented cast irons referred to the determination of the chemical composition and the measurement of the hardness of these alloys. Microstructural analyses were performed by optical metallographic microscopy and by electron scanning-microscopy.

5.1.1.1 Analyses on the chemical composition

Fe - C alloys of cast iron were developed at S.C. Rancon SRL, Iasi Romania. Selected were three types of white cast iron alloyed with chrome (marked A, B, and C) and a grey cast iron with lamellar graphite type FC250 (STAS 568-82), updated under the brand EN - GJL - 250 (SR EN 1561: 1991) and revised under the same name in the technical norm SR EN 1561: 2012. In the study, this cast iron was marked with the letter D (without Cr addition). The melting was performed in induction furnaces and the casting was performed in formation mixtures, based on quartz sand, binder with Kalhartz 8500 resins and Harter hardener [1,2].

For the microstructural characterization of these materials, it was necessary to perform preliminary determination of chemical composition. The chemical composition of the four alloys in the current production of the supplier was determined on the Foundry - Master 01 J 0013 spectrometer from the endowment of the Faculty of Materials Science and Engineering in Iași, Romania. Five measurements were performed on each sample, their average values being presented in Table 5.1.

Table 5.1. Average values of chemical analyses performed.

Chemical composition, [wt%].											
Alloy type	Bookmark	C	Cr	Mn	Si	Mo	Ni	Cu	P	S	rest Fe
White cast iron	A	2.76	11.20	0.65	0.64	0.27	0.30	0.16	0.03	0.07	83.97
White cast iron	B	2.48	14.70	0.53	0.65	0.38	0.38	0.18	0.03	0.07	80.59
White cast iron	C	2.64	20.90	0.66	1.11	0.11	0.16	0.10	0.03	0.05	74.23
Grey cast iron	D	3.10	-	1.00	1.20	-	-	-	0.03	0.04	94.63

The results of the chemical analyses highlight the following aspects:
- the three white cast irons are alloyed with Cr, having different alloying degrees: 11% Cr for cast iron A; 15% Cr - for cast iron B and 21% Cr - for cast iron C;
- white cast irons have similar carbon and manganese content;
- the percentage of silicon is doubles in value for C alloy, compared to the other two white cast iron.

The presence of a medium to high chromium content will exert its influence on the microstructure and properties of white cast irons under study.

5.1.1.2 Hardness measurements

Rockwell hardness measurements were performed for each sample taken from the four cast iron ingots. The tests were performed on a Wilson Wolpert hard drive, model 751H from the Faculty of Materials Science and Engineering, Iasi, Romania. A loading force of 150 kgf was applied on the penetrator for 12s. The average values determined on each cast iron are presented in Table 5.2.

Table 5.2. Rockwell average hardness values.

Alloy type	Bookmark	Determined values for three tests			Calculated average value
		1	2	3	
White cast iron	A	56.7	55.4	56.3	56.1
White cast iron	B	57.3	57.8	56.7	57.3
White cast iron	C	58.9	57.5	58.8	58.4
Grey cast iron	D	23.2	25.8	24.6	24.5

Note: For gray cast iron, the Brinell hardness was actually determined. The average value of 236 HB was subsequently converted to the HRC unit.

The analysis of HRC hardness values highlights the fact that in the presence of high chromium contents considerable quantities of metal carbides will be formed. The increased hardness of these chemical compounds causes hardening of the overall structure of the white cast irons. For non-alloy grey cast iron, the hardness of HRC is kept within normal limits.

5.1.2 Structural characterization of experimental Fe - C alloys

Cast iron for experimental determination belongs to two distinct classes:
- white cast iron highly alloyed with chromium, having % Cr = 10... 21%;
- unalloyed gray cast iron with lamellar graphite.

The properties and operating behavior of these alloys will be influenced by the character of the two main structural components:
- types and quantities of metal carbides formed;
- constitution of the metal mass base (MMB).

The large amount of chromium present in the chemical composition of white cast iron will be distributed as follows:
- the first part will dissolve in cementite, forming a complex alloyed cementite $(Fe,Cr)_3C$;
- the second part will contribute to the formation of specific/own carbides of type M_7C_3 and $M_{23}C_6$, a process directly dependent on the degree of alloying;

- the third part will dissolve in the metal mass base to form solid alloy solutions and mechanical alloy mixtures.

In this way, at ambient temperature, in the structure will be found alloy ferrites (Fα or Fγ) respectively alloyed mechanical mixtures of the type: $E_3 = A + M_7C_3$, for white cast irons with $E_4 = A + M_{23}C_6$ and $E_4 = A + M_{23}C_6$, for white cast irons with Cr> 20% [3].

From the isothermal section, corresponding to the ambient temperature, made in the ternary diagram Fe-C-Cr [4-8], results that at a given content of C, in the structure will be present numerous simple or complex carbides depending on the degree of chromium alloy.

The carbides that form the eutectic colonies from white cast iron alloyed with 8 - 27% Cr can have a fibrous or globulous character and will present:
- lamellar / elongated aspect if the metallographic analysis is performed on a section-oriented parallel to the eutectic growth direction;
- polygonal appearance if the metallographic analysis is performed in a section perpendicular to the direction of growth of the eutectic.

By a hereditary morphology, the allied solid solutions present at ambient temperature will have the appearance corresponding to the austenite specific to high temperatures.

5.1.2.1 Structural characterization performed by light microscopy (optical metallography)

Metallographic analysis was performed on samples from the 4 types of cast iron in the cast state. Because the cast state is not a state of equilibrium, it is more difficult to make a defining comment on the examined microstructures. The investigations were performed on a Meiji Techno -type optical metallographic microscope (Japan) using different magnification powers: 50: 1; 100: 1; 200: 1; 500: 1 and 1000: 1.

The metallographic analysis of white cast iron alloyed with 11% Cr - cast iron A

The investigations performed by optical microscopy on the white cast iron alloyed with 11% Cr - cast iron A highlighted some specific characteristics presented in Fig. 5.1.

The structure consists of grains of solid alloy solution; whose shape corresponds to those of austenite existing at high temperatures. Considering the α-gen character of chromium, the high chromium content and the slow cooling rate when casting in classical forms, there is a tendency to consider that the structure corresponds to an α-alloy ferrite, affected by hereditary morphology (Fig.5.1 - a and Fig. 5.1 - b). At the limits of the grains of solid allied solution, but also inside them, there are complex carbide networks.

Inside some ferritic grains are present the eutectic colonies (Fig. 5.1 - c) in the structure of which are found fine complex carbides.

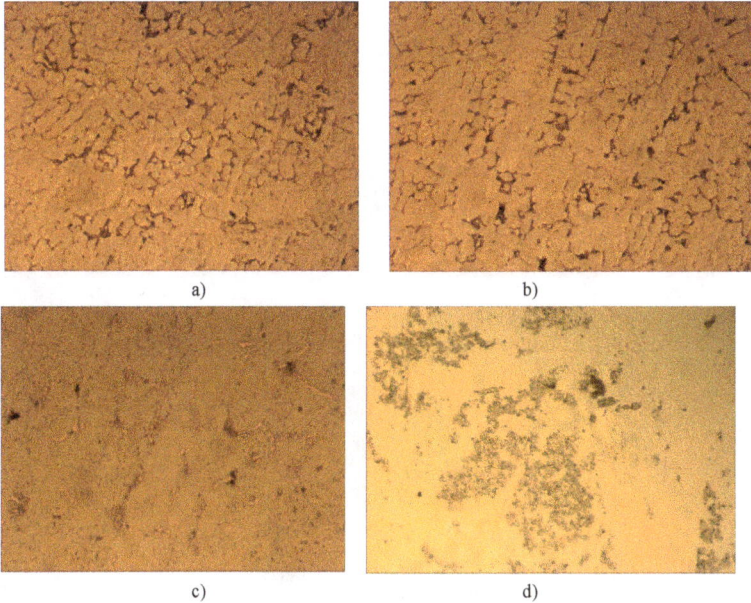

Figure 5.1. Microstructures of white cast iron alloyed with 11% Cr - cast iron A, in cast state, attack with Nital 2.5%: a) magnifying power 50:1; b) magnifying power 100:1; c) magnifying power 200:1; d) magnifying power 500:1.

The dark in color areas correspond to the alloy perlite (Fig. 5.1 - d) formed by fine carbides (similar or different carbides). Fine carbides resulted in the absence of the carbon diffusion process, as they have the character of secondary phases. In the structure are also identified the primary carbides, of large dimensions, favoured by the presence of diffusion.

The metallographic analysis of white cast iron alloyed with 15% Cr - cast iron B

The microstructures identified for white cast iron alloyed with 15% Cr - cast iron B highlight specific features, Fig. 5.2.

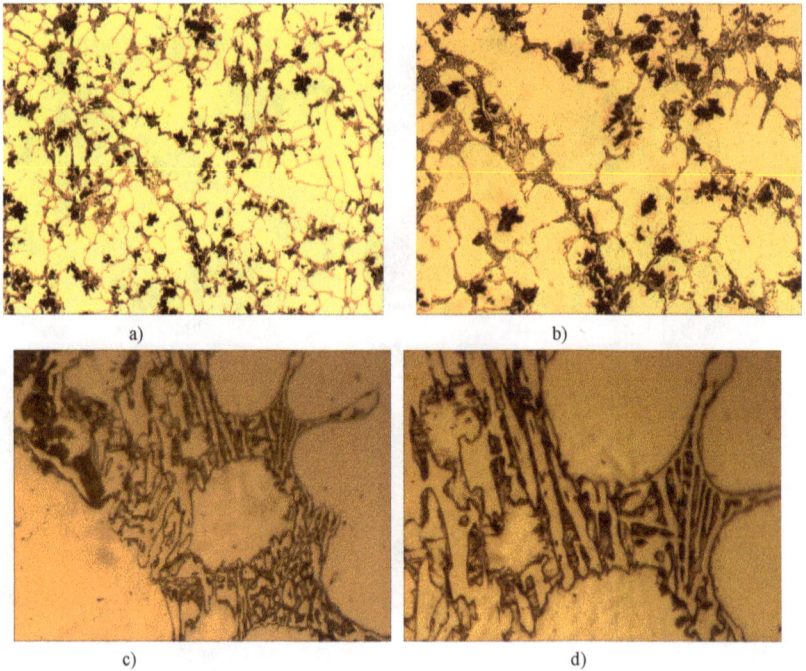

Figure 5.2. Microstructures of white cast iron alloyed with 15% Cr - cast iron B, in cast state, attack with Nital 2.5%. a) magnifying power 50:1; b) magnifying power 100:1; c) magnifying power 500:1; d) magnifying power 1000:1.

The increase in chromium content determines the presence of a larger quantity of solid alloy solution (light colored grains) – Fig. 5.2 - a. eutectic in the constitution of which appear simple or complex carbides of different sizes, Fig. 5.2 - b and Fig. 5.2 - d.

The metallographic analysis of white cast iron alloyed with 21% Cr - C cast iron

In the case of cast iron C, the high chromium content causes a sharp increase in the amounts of solid alloy solution and carbides, Fig. 5.3.

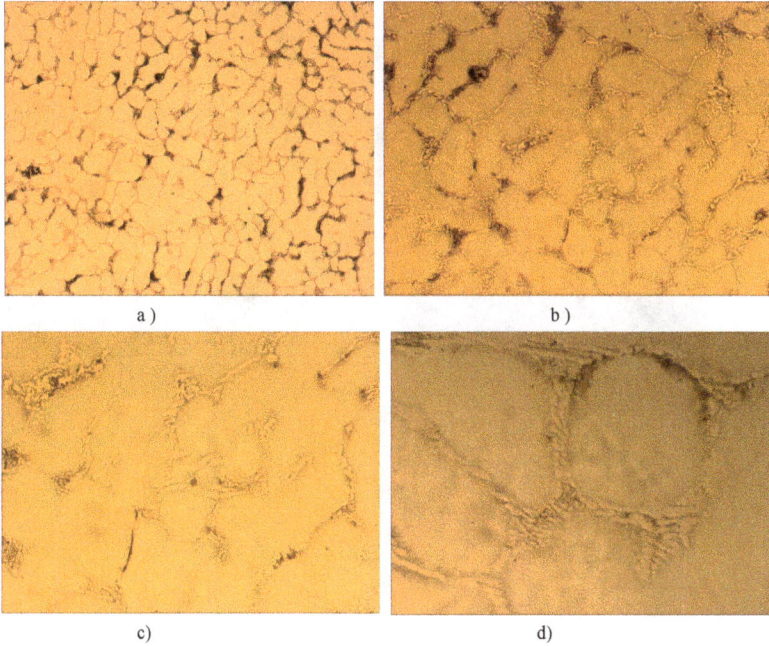

Figure 5.3. Microstructures of white cast iron alloyed with 21% Cr - cast iron C, in cast state, attack with nital 2.5%. a) magnifying power 50:1; b) magnifying power 100:1; c) magnifying power 200:1; d) magnifying power 500:1.

The increase of chromium content determined the accentuated increase of the quantities of solid alloy solution and carbides (Fig. 5.3 - a and Fig. 5.3 - b). In the constitution of eutectic colonies are present complex carbides of different sizes (Fig. 5.3 – c) and Fig. 5.3- d).

The microstructural analysis of grey cast iron with lamellar graphite - cast iron D

This grey cast iron belongs to the class of those with perlito-ferritic structure and lamellar graphite, Fig.5.4.

Without being attacked with chemical reagents, the lamellar character of the graphite inclusions is observed, Fig. 5.4 - a. The graphite has a uniform distribution, the lamellae are thin and with different lengths.

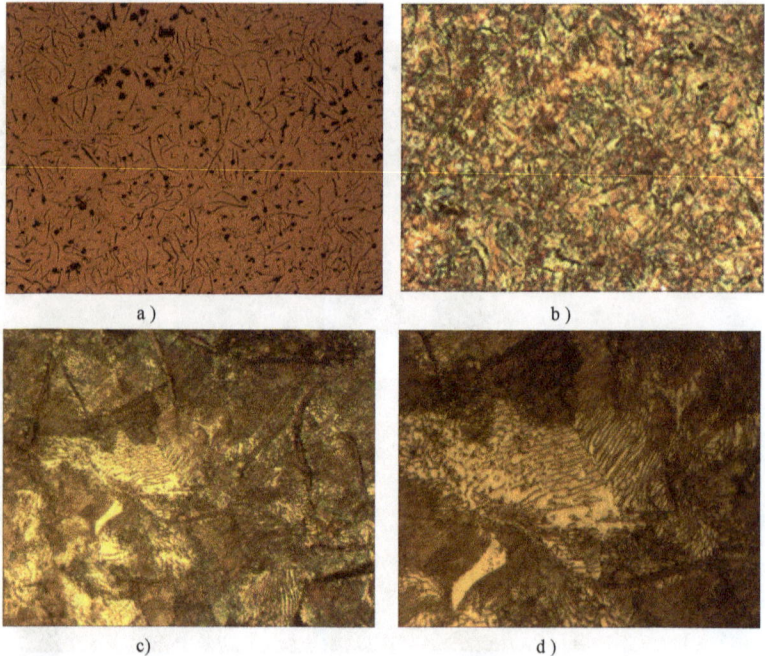

Figure 5.4. Microstructures of gray cast iron with lamellar graphite, in the cast state; a-not chemical etched, b, c and d- attack with Nital 2.5% a) magnifying power 50:1; b) magnifying power 50:1; c) magnifying power 500:1; d) magnifying power 1000:1.

The α ferrite is arranged around the graphite slats, it has the appearance of white areas (Fig. 5.4 - b). Perlite, in large quantities, is in the form of grey islands (Fig. 5.4 - b). The carbides in the perlite constitution have a lamellar character (Fig. 5.4 - c). Large primary carbides are also identified in the structure (Fig.5.4 - d).

5.1.2.2 Structural characterization performed by scanning electron microscopy

For the experimental research performed by SEM electron microscopy, two of the initial Fe-C alloys were selected: white cast iron alloyed with 21% Cr - cast iron C and grey cast iron, D, with lamellar graphite EN-GJL-250.

Analysis on the chemical composition

To achieve structural characterization by electron scanning microscopy (SEM), chemical analysis of the investigated alloy was previously performed.

The experimental determinations were performed on an EDS Bruker equipment equipped with the SEM Vega Tescan LMH II electron microscope (Faculty of Materials Science and Engineering). White cast iron alloyed with 21% Cr - cast iron C was subjected to chemical analysis. On the surface of the sample were selected three points located in areas with different microstructural characteristics (Fig. 5.5 - a), at which chemical composition determination were performed (Fig. 5.5 - b)

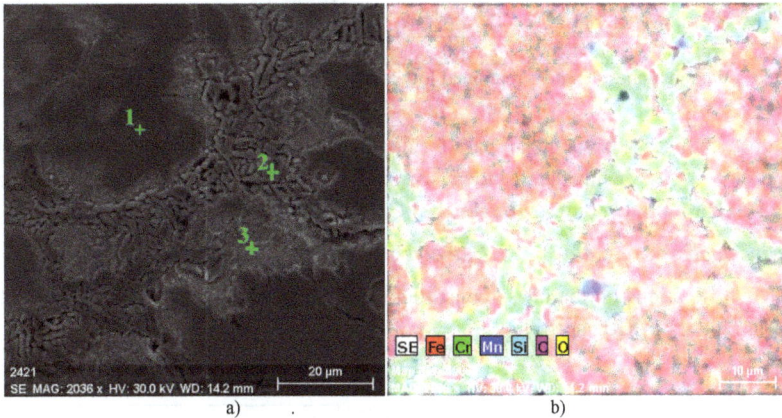

Figure 5.5 Chemical composition measurements by means of EDS a - selection of the points destined for the measurements; b- mapping of the investigated chemical elements (Fe, Cr, Mn, Si, C and O).

The mass and atomic percentage values of the selected chemical elements for white cast iron alloyed with 21% Cr - cast iron 3 are presented in the Table 5.3.

Table 5.3 Percentage values of selected chemical elements for white cast iron alloyed with 21% Cr.

Selected area	Structural component	Chemical Composition									
		Fe		Cr		C		Si		O	
		wt%	at%	wt%	at%	wt%	at%	wt%	at%	wt%	at%
Point 1	Solid solution	76.9	63.5	16.9	14.8	5.1	19.6	1.2	1.9	-	-
Point 2	Double carbide based on iron and chromium	39.7	32.7	55.3	48.9	4,0	15.4	-	-	1.0	3.0
Point 3	Eutectic colony	74.1	64.9	21.5	20.3	2.5	10.0	0.9	0.6	1.1	3.2
EDAX error		1.8		0.5		0.8		0.1		0.6	

Note: As the manganese content remained low at Mn <0.1wt%, it was no longer included in the table

We observe that the basic metal mass formed by solid solution and eutectic colonies have a high iron content, while the carbide phase has a complex character being a double carbide based on chromium and iron type $(Cr, Fe)_7C_3$.

Structural characterization by XRD analysis

Some clarifications on the phase constitution of cast irons under study were possible by determining their XRD spectrum, Fig. 5.6.

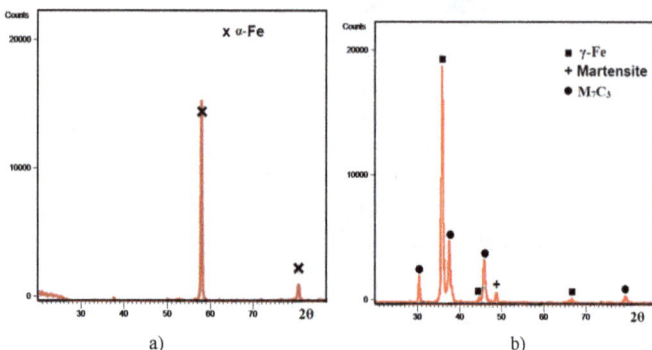

Figure 5.6 Analysis of the phase structure of the studied fonts: a) XRD spectrum of Fe 250 gray cast iron; b) the XRD spectrum of white cast iron alloyed with 21% Cr - C cast iron.

The two XRD diffraction images highlight some structural aspects specific to the analyzed fonts:

- EN-GJL-250 grey cast iron has a large amount of α cast iron and it is of the ferrite-pearlite cast iron type;

- white cast iron alloyed with 21% Cr - cast iron C has a phase structure composed mainly of solid solution (alloyed Fα) and carbides of type M_7C_3. Presence of a small amount of martensite is also noted, resulting from the transformation of residual austenite (austenite that did not transform allotropically into α ferrite upon cooling of the cast iron.

Consequently, with the values of the chemical composition specific to point 2 (Table 5.3.), it is confirmed the existence of the double carbide $(Cr, Fe)_7C_3$ in the structure of the considered cast iron.

5.2   Analysis of ceramic thin layers obtained by plasma spraying

Plasma spray deposits have the advantages of a high deposition rate, large coating areas and the possibility of depositing several materials (metal, ceramic or polymer). For industrial applications, it is important to determine the optimal deposition parameters, which directly depend on the deposited layers and that of the substrate, such as the number of deposited layers, the spray distance or the roughness of the substrate.

5.2.1   Structural and chemical analysis of the metallic material surface after mechanical processing

To cover with a multilayer ceramic material with high properties of corrosion resistance, thermal insulation and properties of increasing the wear resistance of the basic material, cast iron EN-GJL-250, with the surface shown in Fig. 5.7 a) –f), was subsequently machined by blasting.

Three samples were prepared for deposition, one only coarsely ground and cleaned with technical alcohol and ultrasound to remove impurities and grease from the surface, and two by mechanical processing by sandblasting with two degrees of surface deepening. We obtained a surface with a roughness of 0.74 µm, by rough grinding, in which graphite formations are observed [9-12].

Fig. 5.7, c) -d) shows the surface condition of the EN-GJL-250 alloy after blasting. A change in the surface is observed by creating indentations left by the blasting material (roughness of 4.25 µm). The blasting was done on a Rosler equipment intended for dry blasting operations, with glass powder for the lower roughness sample and with aluminum oxide Al_2O_3 Fepa grain 40 (430 µm) on the higher roughness sample, both with a pressure of 3 bar. Fig. 5.7 e) -f) shows the surface condition of the EN-GJL-250 cast iron after mechanical processing of the surface (2-medium regime with a roughness of 2.5 μm) at two image amplification powers of 500x and 2000x respectively.

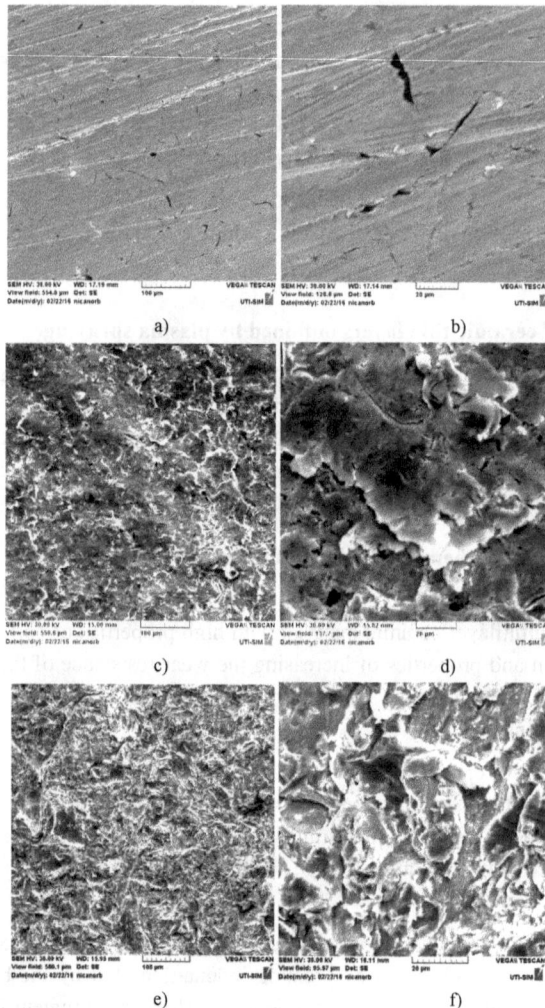

a)

b)

c)

d)

e)

f)

Figure 5.7. The surface condition of the EN-GJL-250 cast iron for two amplification powers of 500 respectively 2000x: a) and b) after sanding; c) and d) glass blasting; e) și f) sandblasting.

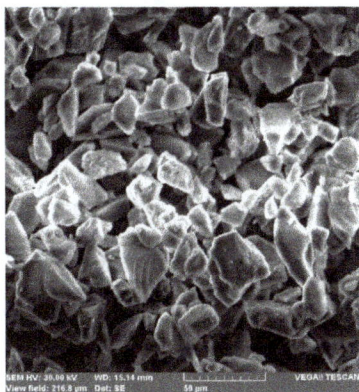

Figure 5.8a. Morphology of the particles that make up the deposition material.

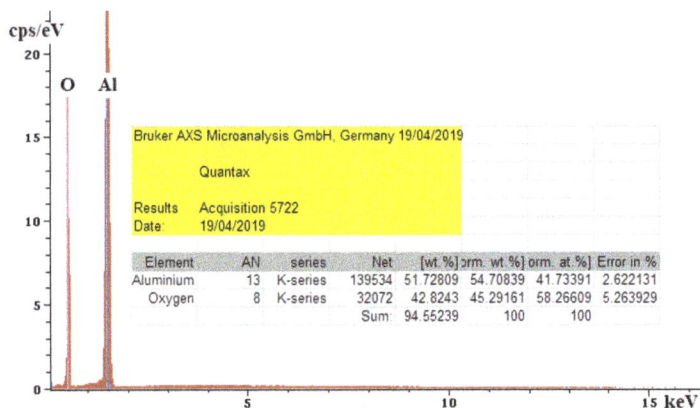

Figure 5.8b. The energy spectrum of the component elements together with their quantitative composition.

After the samples were prepared, one by sanding and two by sandblasting, they were used as a substrate for plasma jet deposition. The deposited raw material, an Al_2O_3 ceramic material has an average particle size of 75 μm.

From the morphology of the particles that make up the powder for deposition layers, Fig. 5.8a), we observe rectangular particles that have minimum dimensions approximately 42μm and 150.71 μm, and a standard deviation of 19.67 μm. Values were determined after 100 particle sizing using the VegaT interpretation program. The quantitative chemical analysis of powders, Fig. 5.8 b) was performed by EDS dispersive energy spectroscopy [13,14].

5.2.2   Structural and chemical analysis of the surface of the metallic material after deposition of the ceramic layer

After the deposition of the complex ceramic material based on Al (alumina) oxides, by passing two or four times with the plasma jet, compact layers were obtained, especially on the sandblasted samples, with estimated thicknesses of 30 and 60 μm, respectively, the thickness of the layer being an approximately linear function with the number of passes [7].

The surfaces of the layers deposited by spraying with "plasma are presented in Fig. 5.9 through scanning electron microscopy. From a structural perspective, the deposits made on the sanded and not sandblasted sample show intermittencies of the layer, Fig. 5.5 a) and b, and areas where the cast iron substrate is observed.

The discontinuities of the deposited layer, of any nature: pores, cracks, exfoliation, etc., worsen the subsequent mechanical and chemical properties. For these reasons, it is not recommended that the plasma jet deposition be performed without prior macroscopic sample preparation to improve the adhesion of the layer.

Fig. 5.9 c) -d) shows the morphology of the layer deposited on glass- sandblasted sample (which has a higher roughness). It is observed that a compact layer has formed on the surface, without interruptions or exfoliation that locally present micro-cracks of ordinal micrometers, Fig. 5.9 d). Cracks appear on the surface without penetrating the entire depth of the layer and are mainly due to temperature differences between the previous layer (made at the penultimate pass), which is cooled and the last layer which is deposited at a high temperature. The layer is composed of areas of molten material (the powders analyzed above) that form a compact mass under the influence of the temperature at which they are brought to values of 12 000 °C. In the last case of surface processing, Fig. 5.9 e) -f), it is observed a homogeneous layer also with areas of molten material and locally with micro-cracks. Microstructurally, it can be appreciated that the ceramic layer with a thickness of approximately 60 μm corresponding to a number of 4 passages has fewer cracks, fewer pores and a more homogeneous coating on the sample with a higher roughness. Higher roughness is also an advantage for the adhesion properties of the ceramic layer because it has a higher anchoring support. The surface of the deposited ceramic layers was also analyzed by atomic force microscopy (2D and 3D) on an area of 64 μm^2, Fig. 5.9 g), and it is also observed in the case of the sample with four deposited layers the achievement of a homogeneous surface.

e)   f)

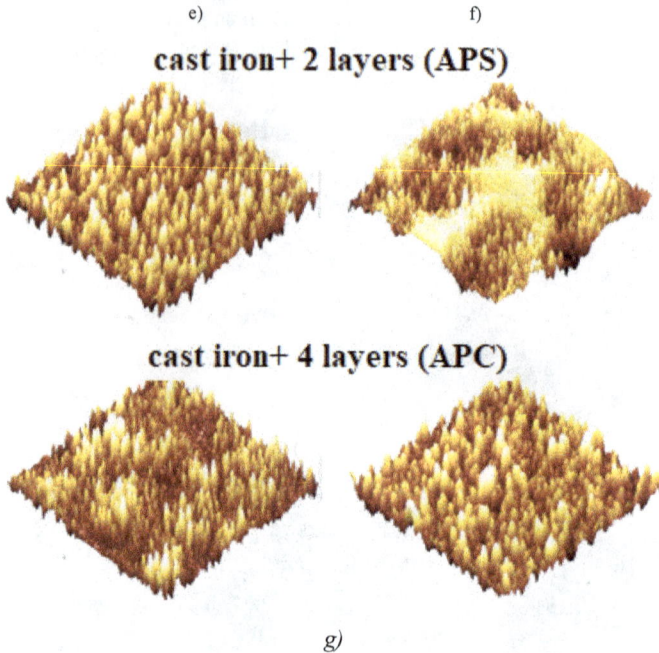

Figure 5.9 SEM and AFM images of the deposited ceramic layer: a) and b) on the sample
without mechanical processing, c) and d) on the sample with glass particle processing; e)
and f) on the sand processing sample; g) AFM images of the deposited ceramic layer.

After establishing a suitable blasting regime, experimental samples were made with two or
four passes of sprayed ceramic material (2 or 4 ceramic layers deposited on the same
sample) using cast iron as substrate.

Fig. 5.10 shows the energy spectrum characteristic of chemical elements qualitatively
identified on the surface of the ceramic layer deposited on cast iron substrate. The presented
spectrum was identified for both experimental samples with two and four deposited
ceramic layers, respectively.

From the analysis of the spectrum in Fig. 5.10 a) it is observed the identification of the
elements characteristic of the ceramic layer respectively Al and O, but also the elements
Fe and C specific to the cast iron substrate. The characteristic elements of the substrate (Fe
and C) were identified through the pores and cracks that appeared in the ceramic layer as

a result of the deposition process. The XRD spectrum, Fig. 5.10 b), identifies several characteristic peaks for the alumina-specific α phase.

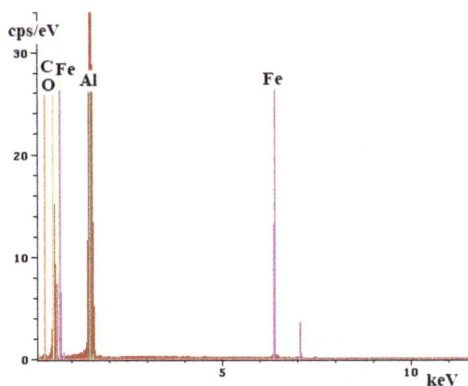

a)

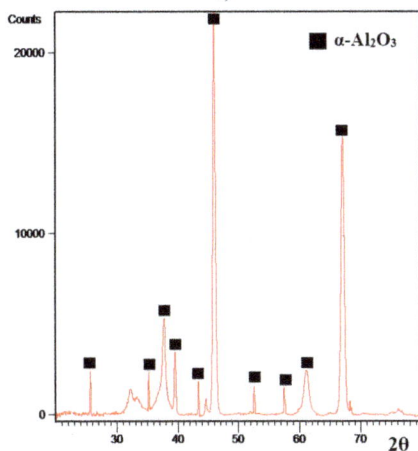

b)

Figure 5.10 Chemical analysis of ceramic layers deposited on metal substrate: a) the EDS spectrum; b) the XRD spectrum.

Table 5.4 shows in mass (wt%) and atomic (at%) percentages chemical composition of the ceramic layers deposited on cast iron substrate. In the case of the sample with two deposited ceramic layers, a high percentage of iron is observed, which means a lower proportion of coating of the metallic material. For the four-pass sample, the identified iron occurs due to micro-cracks or pores existing in the ceramic layer or the reduced thickness of the ceramic layer in some areas which depend on the quality of the deposit, which vary depending on the geometry of the sample. The determination of chemical composition represents an average of the values obtained after 10 determinations, after the analyses being calculated also the standard deviations: Al: ± 1.8, O: ± 0.9, Fe: ± 0.2 and C: ± 0.1. The high percentage of carbon for both cases is due to equipment errors.

Table 5.4 The chemical composition of the ceramic layers deposited on the cast iron substrate.

Element / sample	Al		O		Fe		C	
	wt%	at%	wt%	at%	wt%	at%	wt%	at%
Al_2O_3 powders	54.71	41.73	45.29	58.27	-	-	-	-
EN-GJL-250+2 ceramic layers	40.83	32.72	38.01	49.37	16.5	9.7	4.67	8.21
EN-GJL-250+4 ceramic layers	50.93	38.11	43.78	55.25	1.7	0,6	3.58	6.02
EDS % Error		1.7		1.1		0.2		0.1

For the chemical characterization of the thin layers made by thermal spraying, the distribution of the chemical elements on the ceramic layer deposited on cast iron was followed and the results are presented in Fig. 5.11. The distribution of the following chemical elements was analyzed: Al and O, which make up the deposited ceramic layer and Fe and C, which are characteristic elements of cast iron EN-GJL-250, Fig. 5.11 b) and f). From the analysis of the morphology of the layer and the distribution of the elements Al and Fe, Fig. 5.11c) and d) it is observed in the case of the sample with two deposited layers a high non-uniformity of the deposited ceramic layer covering a little more than 60% of the analyzed surface. of 0.18 mm². The area chosen for the analysis, Fig. 5.11a) is a characteristic of the deposited layer, presenting the same non-uniformities on the entire deposited surface. The non-uniformities of the ceramic layer lead to the total exfoliation of the ceramic layer during use and also to a promotion of the corrosion of the metallic areas exposed to an electrolyte solution compared to the area covered with the ceramic material [15-17].

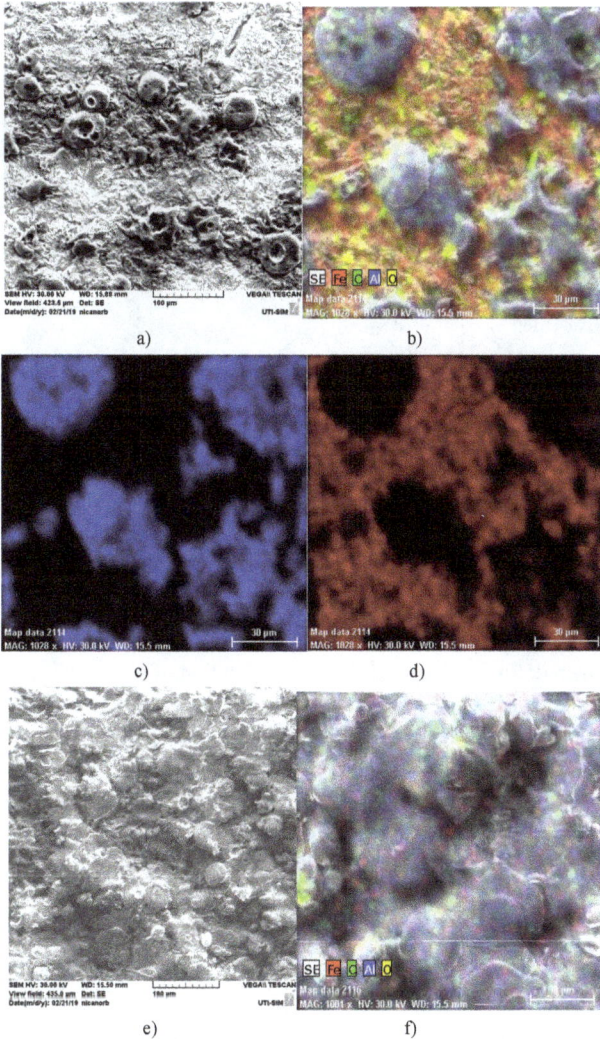

*Figure 5.11A. Distribution of Al, O, Fe and C elements on the deposited ceramic layers:
a) the surface selected on the sample EN-GJL-250 + 2 layers; b) the distribution of the
elements on the surface selected in a); c) Al distribution on the surface selected in a); d)
Fe distribution on the surface selected in a); e) the surface selected on the sample EN-
GJL-250 + 4 layers.*

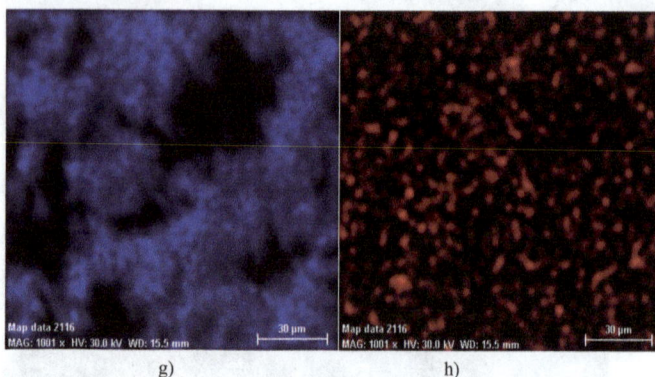

g)   h)

Figure 5.11B. Distribution of Al, O, Fe and C elements on the deposited ceramic layers: g) Al distribution on the surface selected in e); h) Fe distribution on the surface selected in e).

The structural and chemical analysis of the ceramic layer obtained after four passes during the deposition process shows a uniform surface of the layer with all areas of the metal substrate covered, Fig. 5.11 e) and f). The distribution of aluminum, Fig. 5.11 g) and the lack of iron signal, Fig. 5.11 h) confirm that the metal surface is completely covered after deposition with four plasma jet passes. Depending on the application for which the layer is deposited, it can be subjected to additional final rolling operations or heat treatments of chemical and structural homogenization in the furnace or with acetylene flame [18]. Here, the aim is to keep the surface roughness to increase the coefficient of friction, a self-rolling operation occurring in the case of brake discs even during the operation.

References

[1]   N. Cimpoeşu, L. C. Trincă, G. Dascălu, S. Stanciu, S.O. Gurlui, D. Mareci,. Electrochemical Characterization of a New Biodegradable FeMnSi Alloy Coated with Hydroxyapatite-Zirconia by PLD Technique, Journal of Chemistry, (2016) Article ID 9520972. https://doi.org/10.1155/2016/9520972

[2]   C. Florea, C. Bejinariu, C. Munteanu, N. Cimpoesu, Preliminary Results on Complex Ceramic Layers Deposition by Atmospheric Plasma Spraying. Advanced Materials Engineering and Technology V, Book Series: AIP Conference

Proceedings 1835, (2017) Article Number: UNSP 020053.
https://doi.org/10.1063/1.4983793

[3]   L. Sofroni, I. Riposan, I. Chira, Fonte albe rezistente la uzură. Editura Tehnică, Bucuresti 1987.

[4]   I.I. Țîpin, Fonte albe rezistente la uzare, Ed. Metallurghia, Moscova, 1983

[5]   I. Ştirbu, P. Vizureanu, M. Ratoi, N. Cimpoesu, Electrochemical deposition of hydroxyapatite (HA) on titanium alloys for the implant surface bio-functionalization. IEEE E-Health and Bioengineering Conference (EHB) (2013). https://doi.org/10.1109/EHB.2013.6707411

[6]   I. Gradinariu, I. Stirbu, C.A. Gheorghe, N. Cimpoesu, M. Agop, R. Cimpoesu, C. Popa, Chemical properties of hydroxyapatite deposited through electrophoretic process on different sandblasted samples. Mater. Sci.-Poland 32 (2015) 578–582. https://doi.org/10.2478/s13536-014-0241-x

[7]   C.D. Florea, C. Bejinariu, I. Carcea, V. Paleu, D. Chicet, N. Cimpoeşu, Preliminary results on microstructural, chemical and wear analyze of new cast iron with chromium addition, Key Engineering Materials. 660 (2015) 97-102. https://doi.org/10.4028/www.scientific.net/KEM.660.97

[8]   S. Heuer, J. Matějíček, M.Vilémová, M. Koller, Ch. Linsmeier,. Atmospheric plasma spraying of functionally graded steel/tungsten layers for the first wall of future fusion reactors. Surface and Coatings Technology. 366 (2019) 170-178. https://doi.org/10.1016/j.surfcoat.2019.03.017

[9]   D. Delpueyo, X. Balandraud, M. Grediac, S. Stanciu, N. Cimpoesu, Measurement of Mechanical Dissipation in SMAs by Infrared Thermography Residual Stress, Thermomechanics & Infrared Imaging, Hybrid Techniques And Inverse Problems, VOL 9, Edited by:Quinn, S; Balandraud, X, Book Series: Conference Proceedings of the Society for Experimental Mechanics Series, Pages: 9-14, DOI: 10.1007/978-3-319-42255-8_2, 2017. https://doi.org/10.1007/978-3-319-42255-8_2

[10]   J. Izquierdo, G. Bolat, N. Cimpoesu, L.C. Trinca, D. Mareci, R.M. Souto, Electrochemical characterization of pulsed layer deposited hydroxyapatite-zirconia layers on Ti-21Nb-15Ta-6Zr alloy for biomedical application, Applied Surface Science 385 (2016) 368-378. https://doi.org/10.1016/j.apsusc.2016.05.130

[11]   D. Delpueyo, X. Balandraud, M. Grediac, S. Stanciu, N. Cimpoesu, A specific device for enhanced measurement of mechanical dissipation in specimens

subjected to long-term tensile tests in fatigue, Strain 54 (2018)Article Number:e12252. https://doi.org/10.1111/str.12252

[12] R. Kromer, S. Costil, C. Verdy, S. Gojon, H. Liao, Laser surface texturing to enhance adhesion bond strength of spray coatings – Cold spraying, wire-arc spraying, and atmospheric plasma spraying. Surface and Coatings Technology. 352 (2018) 642-653. https://doi.org/10.1016/j.surfcoat.2017.05.007

[13] A.P. Markopoulos, N.E. Karkalos, M. Mia, D.Y. Pimenov, M.K. Gupta, H. Hegab, N. Khanna, V.A. Balogun, S. Sharma, Sustainability Assessment, Investigations, and Modelling of Slot Milling Characteristics in Eco-Benign Machining of Hardened Steel, Metals 10 (2020)Article Number: 1650. https://doi.org/10.3390/met10121650

[14] X. Balandraud, X.; J. –B. Le Cam, Some specific features and consequences of the thermal response of rubber under cyclic mechanical loading, Archive of Applied Mechanics 84 (2014) 773-788. https://doi.org/10.1007/s00419-014-0832-3

[15] A. Mekaouche, F. Chapelle, X. Balandraud, Using shape memory alloys to obtain variable compliance maps of a flexible structure: concept and modeling, Meccanica 51 (2016) 1287-1299. https://doi.org/10.1007/s11012-015-0301-2

[16] X. Balandraud, N. Barrera, P. Biscari, M. Grediac, G. Zanzotto, Strain intermittency in shape-memory alloys, Physical Review B, 91 (2015) Article Number: 174111. https://doi.org/10.1103/PhysRevB.91.174111

[17] V. Paleu, G. Gurau, R.I. Comaneci, V. Sampath, C. Gurau, L.G. Bujoreanu, A new application of Fe-28Mn-6Si-5Cr (mass%) shape memory alloy, for self-adjustable axial preloading of ball bearings, Smart Materials and Structures, 27 Article Number: (2018) 075026. https://doi.org/10.1088/1361-665X/aac4c5

[18] A.P. Markopoulos, N.E. Karkalos, E.L. Papazoglou, Meshless Methods for the Simulation of Machining and Micro-machining: A Review, Archives of computational methods in engineering 27 (2020) 831-853. https://doi.org/10.1007/s11831-019-09333-z

Chapter 6

Mechanical Characterization of the Materials for Automotive Brake Systems

C.D. Florea[1], N. Cimpoesu[1], C. Bejinariu[1], R. Cimpoesu[1]*

[1]Faculty of Materials Science and Engineering, "Gheorghe Asachi" Technical University of Iasi, Romania

ramona.cimpoesu@tuiasi.ro

Abstract

In the case of materials proposed for applications in which friction is involved, the characterization of their tribological behaviour is essential. For the analysis of the wear behavior of some materials, regardless of their nature, the profilometry of the material surface and the characteristics of the proposed experiment are taken into account.

Keywords

Tribological, Wear, Profilometry, Scratch

6.1 Profilometry of EN-GJL-250 material and ceramic layers

The profilometry measurements of the experimental samples with deposited layers but also of the initial substrate were performed using the Taylor Hobson profilometer. Several types of surfaces were measured to highlight the distribution of material and the way the profiles look for different cases.

The specimens were fixed on the universal table and over them passing with the tip's arm, which moved at a constant speed over the sample. The probe detects the surface deviations using the transducer. It produced an analogue signal that matched the probe's vertical movement. This signal was then amplified, sampled, quantified and displayed by a two-dimensional map of the rough surface. Fig. 6.1 shows a profile of a cast iron surface, the sample being taken from a worn brake disc and changed after 7 years of operation in heavy conditions. We can analyze the aspect of the surface profile that has the material distributed in a Gaussian shape. The values for the parameters skewness and kurtosis are approximate with the values corresponding to the Gaussian distribution (Sk = 0 and K=3). One can observe the symmetrical distribution of the material with respect to the midline and fact that this profile has the number of peaks approximately equal to that of the depressions.

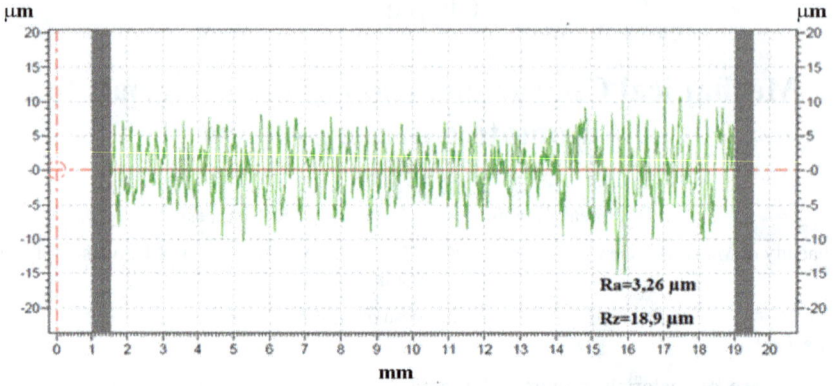

a)

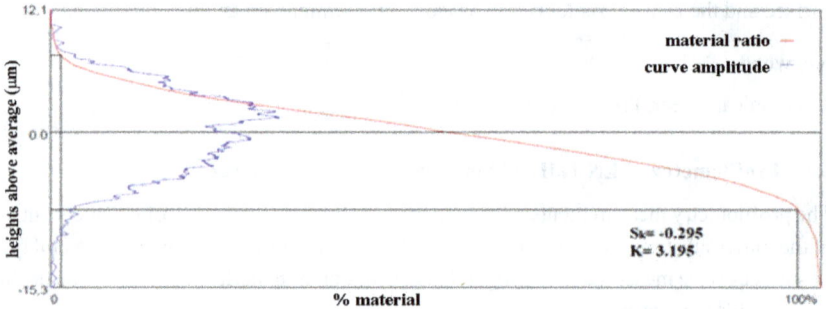

b)

Figure 6.1. Initial cast iron sample profile, EN-GJL-250: a) surface condition; b) above average height distribution.

Usually, the roughness of a surface appreciates the variation in the height of the real surface in relation to a nominal surface. The roughness can be measured along a single surface profile obtaining a two-dimensional characterization (2D) or along a set of parallel profiles, obtaining a three-dimensional characterization (3D). The value of the skewness parameter depends on how the solid material is distributed within the profile, relative to the mean line [1-3]. If it is above the mean line then the value of the skewness parameter is negative, Fig.

6.1 b), and if the solid material is below midline then the value of the skewness parameter is positive.

A symmetrical distribution of heights for which the number of peaks is equal to that of the depressions determines that the parameter Sk is approximately equal to 0. Thus, the skewness parameter can be used to differentiate surfaces that have the same value for the arithmetic mean height although they are characterized by different shapes of the profiles.

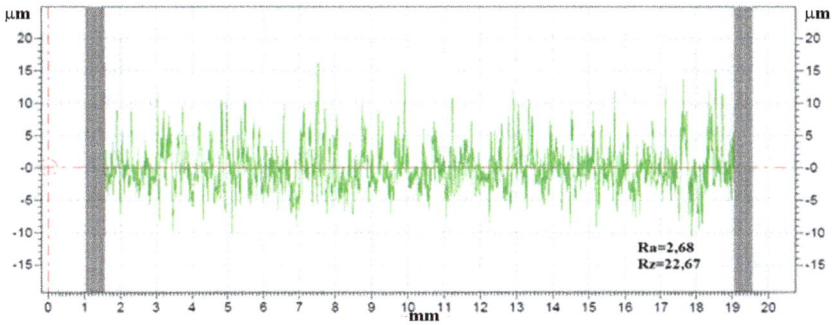

a)

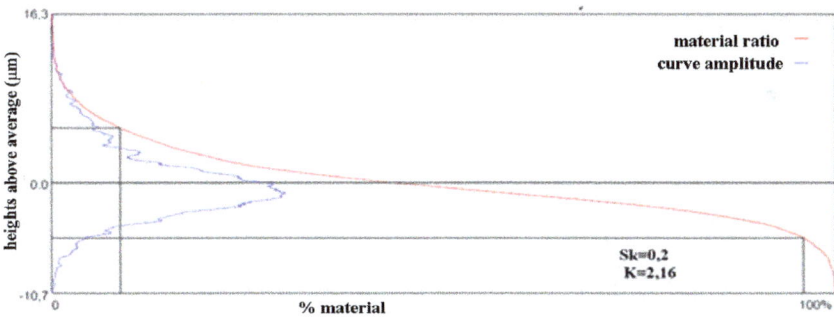

b)

Figure 6.2. Analysis of the surface profile of the cast iron sample with two layers of ceramic material: a) the condition of the surface; b) above average height distribution.

The kurtosis parameter also appreciates the shape of the profile. Thus, if within the limits of the reference length the profile has relatively few high peaks and recesses, Fig. 6.1b), results k<3 (platykurtic profile), and if in contrast the profile has many high peaks and recesses results k>3(profile leptokurtic) [4,5].

The height of the roughness at a point on the abscissa X of the profile is a random variable, being characterized by the statistical set of its values, in all possible profiles. In the following processes, the roughness height will be considered as an ergodic stationary random variable, the average values defined for a profile not depending on the chosen profile and coinciding with the average values defined as a whole. In the case of sample with two ceramic layers, we highlight a profile with kurtosis parameter value less than 3. The profile of the cast iron specimen on which two layers of Al_2O_3 ceramic material were deposited by thermal spraying is presented in Fig. 6.2 a) -b). It can be seen that the sample profile has relatively few high peaks and recesses which cause flattening of the distribution curve. Fig. 6.3 shows a profile, the sample with four deposited Al_2O_3 ceramic layers, which is characterized by the following values for the skewness parameters Sk = 0.08 and kurtosis close to the value 3.

A profile that has the positive skewness parameter is a profile in which the tips have not been removed and most of the solid material is distributed above the midline. A slight shift to the right of the line of symmetry of the distribution curve can be observed. As we have seen before, the skewness parameter is useful in evaluating the asymmetry of the roughness height distribution in relation to the mean line.

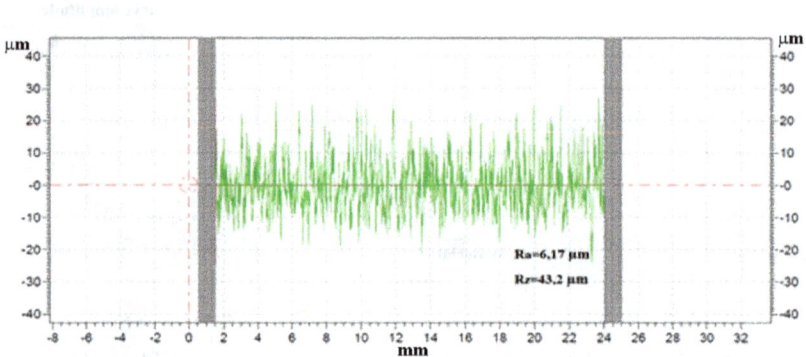

a)

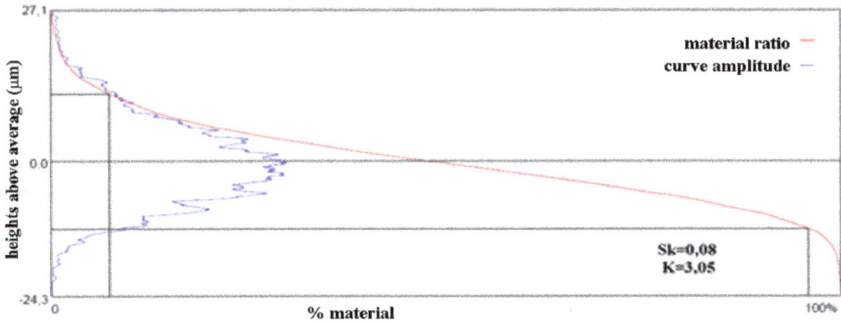

b)

Figure 6.3 Analysis of the surface profile of the cast iron sample with four layers of ceramic material: a) the condition of the surface; b) above average height distribution.

The symmetric and the Gaussian distribution have the value zero for the skewness parameter, while the asymmetric distribution in relation to the mean line of the roughness height also determines an asymmetric shape of the probability density function, Fig. 6.4. The Gaussian distribution of roughness heights has the value $k = 3$. It has been shown that the size of the kurtosis factor provides indications on the shape of the roughness profile. It is considered an orthogonal system having in abscissa, the value of the skewness parameter and in ordinate the value of the non-normalized kurtosis factor, Fig. 6.4.

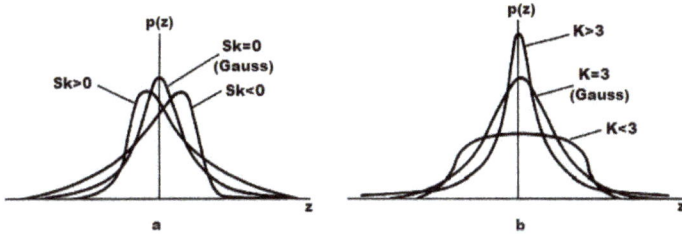

Figure 6.4 Orthogonal system having in abscissa the value of the Skewness parameter and in ordinate the value of the non-normalized kurtosis factor: a) analysis of the Sk parameter; b) analysis of parameter K.

Although there is a tendency to assimilate a priori the surfaces of machine parts as having a Gaussian-type roughness distribution, practical applications clearly indicate that many of the manufacturing technologies lead to surfaces where microtopography is far from Gaussian preferred for the control of the behavior of some experimental samples.[6]

6.2 Analysis of the behavior of experimental materials with ceramic layers at micro-indentation

It is considered that the preparation of test surfaces using mechanical methods is a possible cause for the increase in the apparent hardness and the decrease of the indentation size. These are attributed to the deformation of the surface layers by the preparation process. Such statements are true in the preparation process but do not make sense as a generalization unless the details of the preparation procedures are known and have been assimilated.

In the same sense, it is assumed that an electrically polished surface is intrinsically devoid of the effects of these imperfections and can be considered as unaffected case. It can also be true or false depending on the details of abrasion and processing used before and during polishing. Here, the experimental samples did not undergo surface processing, being analyzed in the state in which they were obtained, respectively, after cutting the initial sample and after depositing the samples with ceramic surface layers. Fig. 6.5 presents the behavior of experimental materials at micro-indentation. Fig. 6.5a) gives the variation graphic of the pressing force and Fig. 6.5 b) the variation of the coefficient of friction as a function of time.

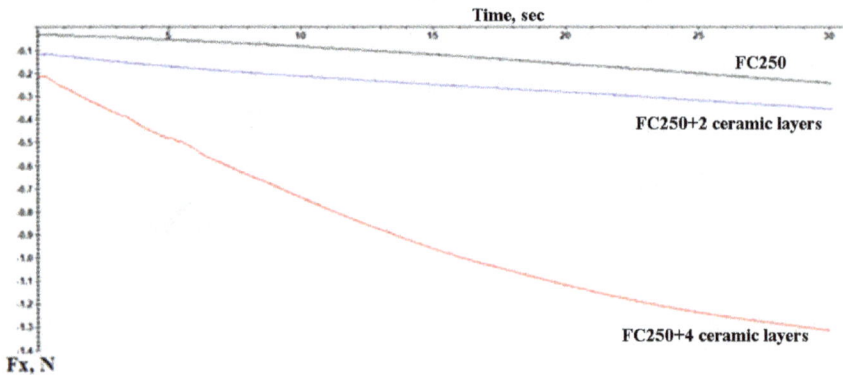

a)

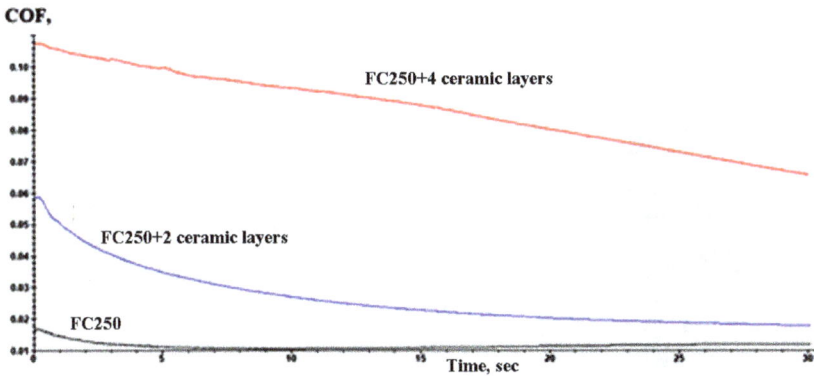

Figure 6.5 Behavior of experimental materials at micro-indentation: a) variation of the pressing force; b) variation of the coefficient of friction as a function of time.

Both variations, penetration force and coefficient of friction, shown in Fig. 6.6 a) and b), show a hardening of the surface by depositing ceramic surface layers with an increase of the coefficient of friction up to five times, recorded during the test penetration. In comparison, the deposited two-layer sample behaves more like the substrate than the deposited four-layer ceramic sample. The fluctuations that occur at 180 and 330 seconds, respectively, in the evolution of the pressing force and in the coefficient of friction of the sample with four deposited layers, can be attributed to the penetration of the last two deposited ceramic layers. For the analysis of the ceramic layer's adhesion to the mechanical substrate and also for the determination of the tribological properties, scratch tests were performed on the experimental samples: EN-GJL-250, EN-GJL-250 + 2 ceramic layers and EN-GJL-250 + 4 ceramic layers. Fig. 6.6 shows the general appearance of the scratch marks from the beginning of the test, from the left to the end on the right. The compound figures were made from three optical microscopy of the areas that characterize the scratch from the beginning of the test to the end. The length of the scratches was 25 mm with an increasing pressing force. Several scratch tests were performed to characterize the homogeneity of the tribological properties of the deposited surface layers [7-10].

At the macroscopic level, no exfoliation of the deposited ceramic layer is observed and their uniform appearance shows a good structural homogeneity of the ceramic layer. Fig. 6.6 shows also the optical aspect at scratching of the EN-GJL-250 sample + 4 ceramic layers. The scratching equipment worked at the same time with an acoustic sensor to record the behavior of the ceramic layer, initially, and the ceramic layer pierced together with the substrate after some time [1, 11].

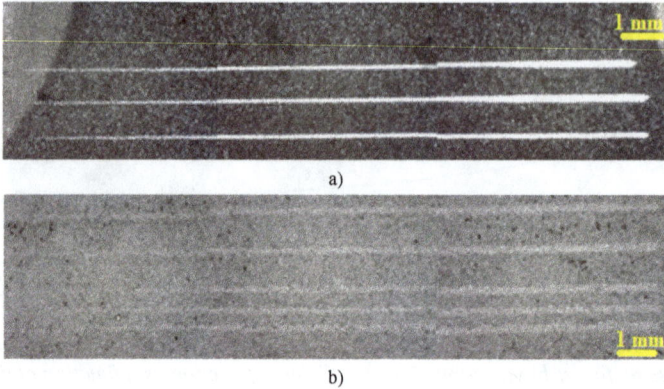

Figure 6.6 Identification and appearance of traces of scratch by optical microscopy: a) 2 layers; b) 4 layers.

The experiment was performed starting from an initial pressing force of 0N (Fz) to 8N over a length of 25 mm. Fig. 6.7 shows the evolution of the friction force (Fx) and acoustic emission (AE) over a distance of 25 mm travelled in 25 seconds (scratching speed being 1 mm / second) [12-14].

It is observed in the evolution of the friction force but also of acoustic emission, a variation of the signals at 10.5 - 11.5 mm from the beginning of the scratching of the ceramic surface, the area that represents the place where the metal penetrator affected the ceramic layer of material. The friction force further increases due to the double stress input, in addition to the initial ceramic layer being opposed the metal substrate EN-GJL-250. For the analysis of the friction coefficient behavior, Fig. 6.8, extracted from the scratch test signal, the scratch made on the ceramic layer and subsequently after its penetration on the EN-GJL-250 ceramic layer-cast iron system is also chemically analyzed on various areas.

The coefficient of friction shows a behavior similar to the friction force with a variation after 10.5 seconds from the beginning of the test, Fig. 6.8 a). The scratch left by the mechanical test was analyzed by SEM electron microscopy (on areas a) to g)) after characterizing the trace every 2 mm, Fig. 6.8.

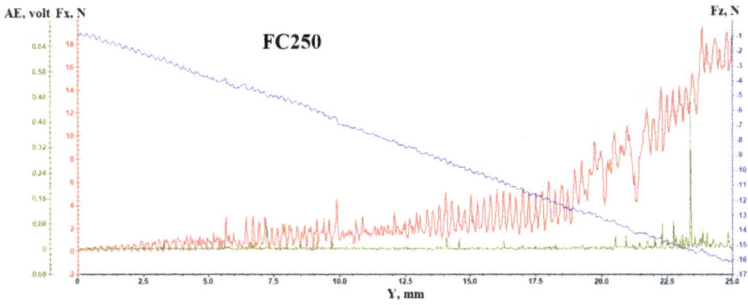

a)

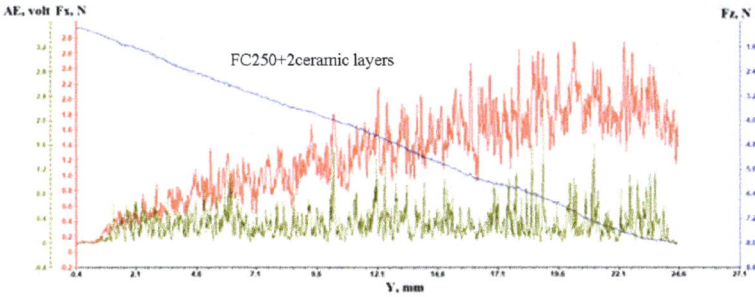

b)

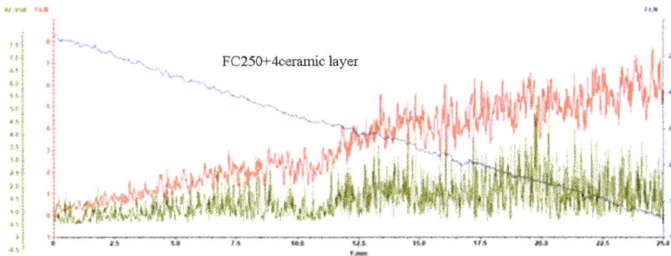

c)

Figure 6.7 Characteristics of scratch behavior: a) the sample from EN-GJL-250; b) sample from EN-GJL-250 + 2 ceramic layers; c) sample from EN-GJL-250 + 4 ceramic layers

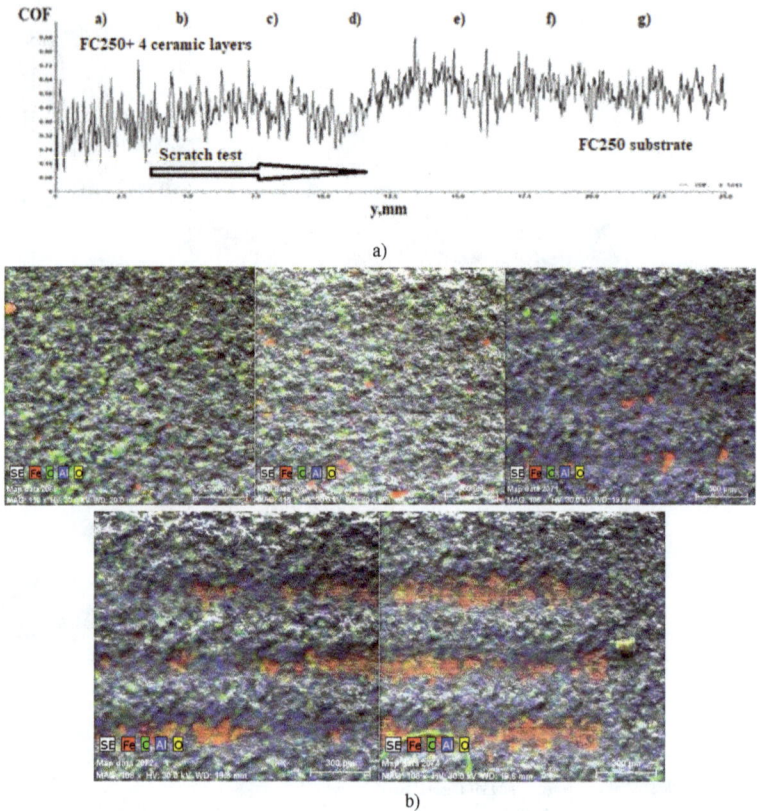

Figure 6.8 Analysis of the scratch test by the behavior of the coefficient of friction: a) variation of the coefficient of friction over the distance of 25 mm; b) the distribution of the elements Fe, C, Al and O on the surface of the scratched ceramic layer (for area a), b) d), f) and g) in figure a)).

Fig. 6.8 b) shows the distribution of the elements Fe, C characteristic of the cast iron substrate and Al, O characteristic of the Al$_2$O$_3$ ceramic multilayer on areas a), b), d), f) and g) from the scratch test. In the first two distributions in Fig. 6.8 b) no penetration of the ceramic layer is observed, this process being highlighted in area d) by the significant increase of the signal of the iron element on the scratch marks. The signal of the iron element is accompanied, less obviously due to the much smaller percentage, by the signal of the carbon element. If in zone d) the ceramic layer was only partially pierced, in the next 10-14 mm it was gradually removed completely in some areas, especially on the last part of the load. You can see in areas f) and g) in Fig. 6.8 b) parts with the ceramic layer present on the scratch marks.

Their presence can be explained by a superior adhesion to the substrate in these areas or by compaction of the ceramic material under the scratching/pressing force and penetration of the EN-GJL-250 metal matrix.

Fig. 6.9 shows the surface condition of approximately 4 mm^2 in different areas of the scratch mark (a) -g) in Fig. 6.9 a). The microstructural analysis was performed starting with the final end of the scratch every 2 mm until no microstructural variations of the ceramic layer were observed, an area considered at the beginning of the scratch test and which corresponded to the area obtained from the calculations, taking into account the length of the trace, respectively 25 mm of scratch. Macro-structurally, a chamfering of the ceramic layer is observed from 2 mm compared to the beginning of the scratch test, i.e. at a force of $1 \div 2$ N, which confirms that the Al$_2$O$_3$ layers are relatively soft between the ceramic materials but less brittle compared to the very hard layers [15]. The deepening of the scratch marks without penetrating the ceramic layer continues until $10 \div 11$ seconds of solicitation, at $10 \div 11$ mm from the beginning of the test, with a force of ~5N. No areas with macro-cracks are observed on the edge of the scratch marks nor in the areas of ceramic material between the scratches, Fig. 6.9 g). Analysis at a higher amplification power of the surface image, Fig. 6.10 a) -d) did not show cracks or pores on the compacted ceramic surface nor their appearance on the metal substrate.

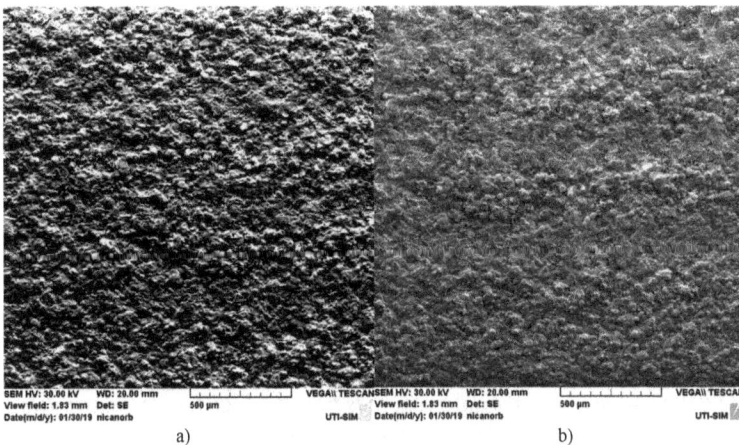

a)   b)

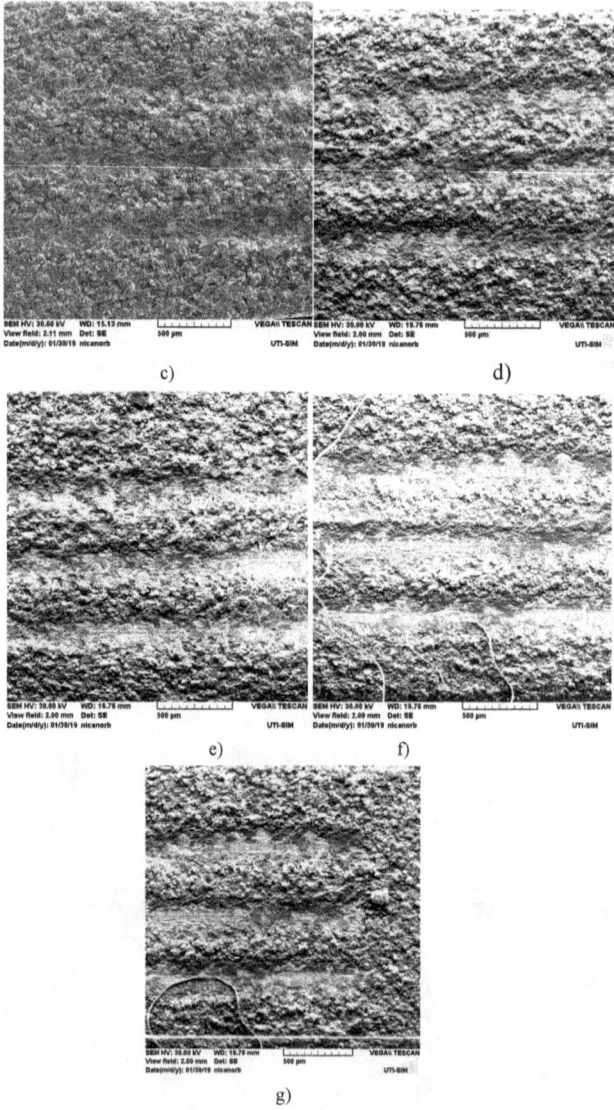

Figure 6.9 SEM images of the different areas from the beginning of the scratch, (a) to the end of the scratch (g).

The integrity of the ceramic layer is little affected on the edges of the scratch mark, which shows a high stability of the ceramic layer. Following the analyzed scratch, there are areas of exfoliation of the ceramic layer, Fig. 6.10 c) but also the presence of areas with compressed ceramic layer. In practical applications where the increase in the wear coefficient is also not pursued, mechanical or thermal processing of the deposited layer is recommended to smooth the surface, reducing the roughness and homogenize the coatings.

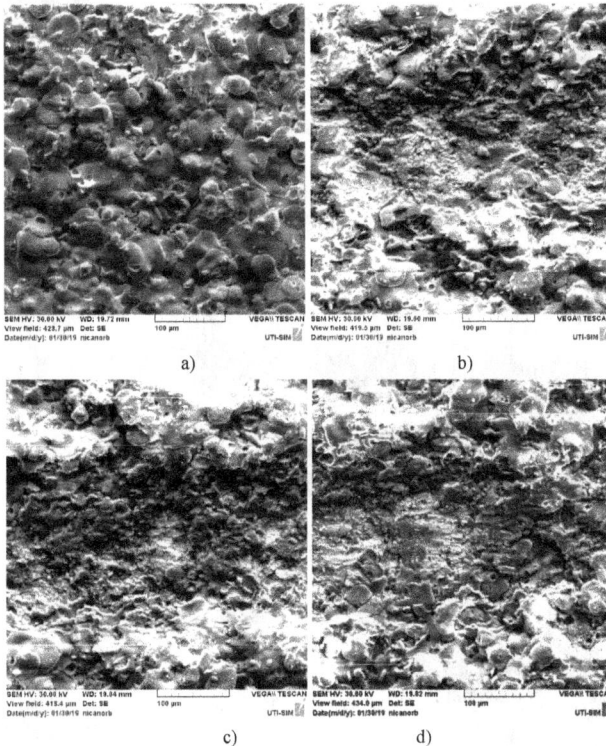

a)   b)

c)   d)

Figure 6.10 SEM images of the details of the areas with the scratching of the ceramic surfaces.

Fig. 6.11 shows the behavior of the experimental materials EN-GJL-250, EN-GJL-250 + 2 ceramic layers and EN-GJL-250 + 4 ceramic layers at the scratch for the variation of a) friction force, b) acoustic emission and b) of the coefficient of friction during the scratch test.

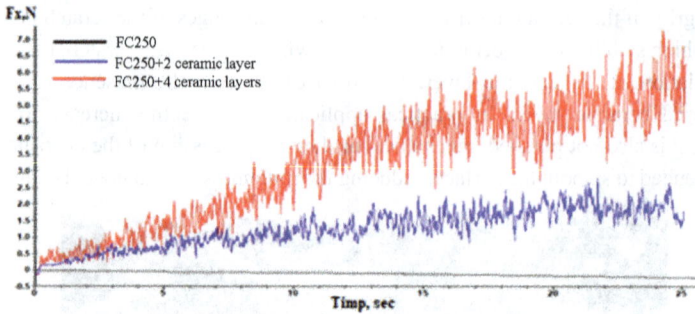

a)

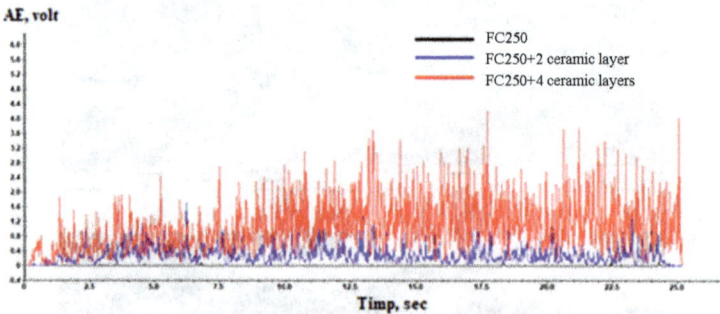

b)

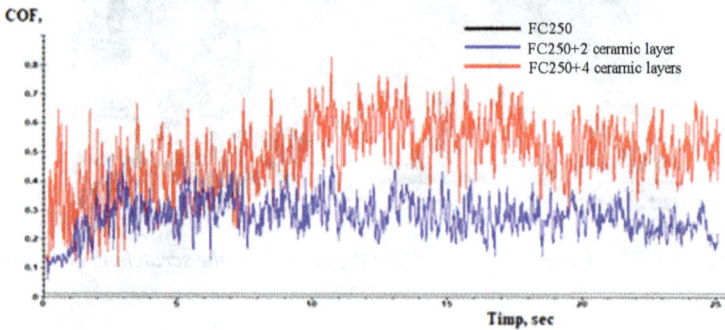

c)

Figure 6.11 Behavior of experimental materials EN-GJL-250, EN-GJL-250 + 2 ceramic layers and EN-GJL-250 + 4 ceramic layers when scratched: a) friction force; b) acoustic emission; c) the coefficient of friction.

Fig. 6.11a) shows that the friction forces are higher in the case of samples with ceramic deposits compared to the friction force that appears on the EN-GJL-250 material and which shows only small variations in behavior due to differences in hardness between the metal matrix characteristic of graphite castings and formations. In both cases with deposits (with 2 and 4 layers respectively) an increase in the friction force is observed after the penetration of the ceramic layer and the complex friction between the indenter on the one hand and the ceramic layer and the metallic substrate on the other. It is also observed a 2 - 3 times increase in the friction force in the case of the sample with 4 ceramic layers deposited compared to the sample with 2 ceramic layers. In the case of acoustic emission (AE), Fig. 6.11 b), the substrate signal is also a straight line compared to the emission of ceramic layer samples.

The creation and propagation of cracks can be events with short occurrence and growth times . Acoustic emissions are designed to detect the behavior of fracturing and cracking of materials. Wakayama and Ishiwata [5] used AE detection to analyze the evaluation of ceramic micro-cracks, to detect damage in ceramic composites and fiber-reinforced metal matrix composites and used it to detect cracking of the work piece during the surface-processing process of engineering ceramics [16, 17].

In the scratch test, the AE technique was also used to monitor fragile rupture events [8]. The level of acoustic emission is higher in the case of the sample with 4 ceramic layers having a visible increase in the area where the deposited layer was penetrated. The amplitude of the acoustic emission signal has increased significantly due to severe indenter vibrations resulting from the initiation or propagation of cracks and / or removal of material by plastic deformation or brittle cracking of the Al_2O_3 layer during scratches. In the case of large fluctuations of the AE signal, the larger the magnitude, the more serious was the damage caused to the ceramic layer or the metal substrate in the areas where it was reached.

The coefficient of friction, Fig. 6.11 c), shows a substantial increase in the case of samples with ceramic layers compared to the cast iron substrate. The increase is due to both the roughness of the ceramic layers and their nature. After penetration of the ceramic layer, the friction of the metal substrate is added to its coefficient of friction, which thus contributes to the increase of the values of the coefficient of friction - COF which vary over a wide range due to the high resistance of the ceramic layer, determining a higher plastic deformation capacity than the substrate. Despite this relatively high coefficient of friction-COF values, plasma spray coatings made of ceramic materials are still accepted for many applications [18]. For applications in severe wear conditions, these coatings can be supplemented by various additional treatments such as laser re-modelling, sealing treatment or surface grinding to improve the modification of the surface and therefore the values of the coefficient of friction - COF [11, 12]. If we follow the evolution of the friction coefficients of the ceramic materials in the form of layers until their penetration (in the time interval 9 - 11) from 0 to 10 seconds of the test, we can observe two areas of variation of the friction coefficient in the figure of the coefficient of friction.

Initially, the coefficient of friction increases sharply to 0.6 with a period of stability, followed by a slight decrease to 0.3 - 0.4 due to the higher degree of flattening of the deposited ceramic layer and lower surface roughness for these coatings. According to the Czihos model [13], the curves of the coefficient of friction - COF consist of three stages: the initial wear, the state of equilibrium and acceleration at wear to the point of breakthrough and contact with the metal substrate in this case. The third step in the friction coefficient curves - COF for the covered and worn samples in this study is opposed to the Czihos model. Some tribological interactions due in particular to the common influence of the substrate and the ceramic edges of the penetrated layer affect the stabilization of the coefficient of friction - COF.

6.3 Experimental determination of the wear resistance of friction alloys

The experimental samples, chromium cast irons and cast iron coated with ceramic layers, were analyzed for wear resistance using Amsler laboratory equipment with a steel friction roller.

6.3.1 Analysis of the wear behavior of experimental samples with chromium additions

Before each test, a solvent model SF D500 was used to clean the samples and the effect of this solvent disappears in two minutes or less depending on the material used. Samples with additions of chromium: A and C were analyzed in comparison with standard cast iron EN-GJL-250 (D).

Three values of the characteristic coefficient of friction were obtained: the average value during all tests, in μm, presented in Table 6.1; the highest value during all tests, μ_{max}, presented in Table 6.2 and the average value of the coefficient at the end of the test, μ_{min}, presented in Table 6.3.

Generally, the average values of the coefficient of friction include the beginning of the test when the friction is reduced to low values of roughness. For the values corresponding to the maximum coefficient of friction, Fig. 6.12, and for the average coefficient of friction at the end of the tests, Fig. 6.13, time was also recorded at various intervals. At low speeds, 50 rpm and 100 rpm, maximum value of the coefficients of friction are at the end of the test, but at 200 rpm and 250 rpm these values and even the realization time are different except for the C sample test at 250 rpm.

Table 6.1 Average coefficient of friction, rpm, during tests for samples A, C and D.

Sample	Average coefficient of friction during tests (15 minutes period)			
	50 [rpm]	100 [rpm]	200 [rpm]	250 [rpm]
A	0.085	0.073	0.071	0.132
C	0.109	0.114	0.140	0.106
D	0.068	0.058	0.104	0.113

Table 6.2 Maximum coefficients of friction during tests and depending on speed [rpm].

Sample	Maximum coefficients of friction during tests and depending on speed [rpm]							
	50 [rpm]	Time [s]	100 [rpm]	Time [s]	200 [rpm]	Time [s]	250 [rpm]	Time [s]
A	0.112	700-900	0.100	700-900	0.095	450-650	0.151	100-250
C	0.145	700-900	0.128	600-900	0.149	100-150	0.134	850-900
D	0.084	750-900	0.071	750-900	0.117	350-450	0.119	125-250

Table 6.3 Average coefficients of friction at the end of wear tests (after 15 minutes).

Sample	Average coefficients of friction at the end of wear tests (after 15 minutes)							
	50 [rpm]	Time [s]	100 [rpm]	Time [s]	200 [rpm]	Time [s]	250 [rpm]	Time [s]
A	0,112	700-900	0.100	700-900	0.084	750-900	0.128	700-900
C	0,145	700-900	0.128	600-900	0.136	600-900	0.134	850-900
D	0.084	750-900	0.071	750-900	0.116	600-900	0.117	700-900

Based on the experimental results obtained if we consider a linear dependence between the coefficient of friction and speed the highest coefficient of friction is obtained for sample C. In the case of sample C, the coefficient of friction is not only the highest but has an independence between speed and time (Fig. 6.12), and compared to sample D (classical cast iron EN-GJL-250) and sample A (cast iron with a lower Cr content of the three chromium cast irons) it has a higher coefficient of friction.

The values obtained for the coefficient of friction correspond to a mixed friction regime and the wear particles trapped in the contact area between the two metallic materials act as a solid lubricant decreasing in friction at low speeds where the speed of removal of abrasive particles is diminished. Fig. 6.13 shows the variation between the average power loss and speed.

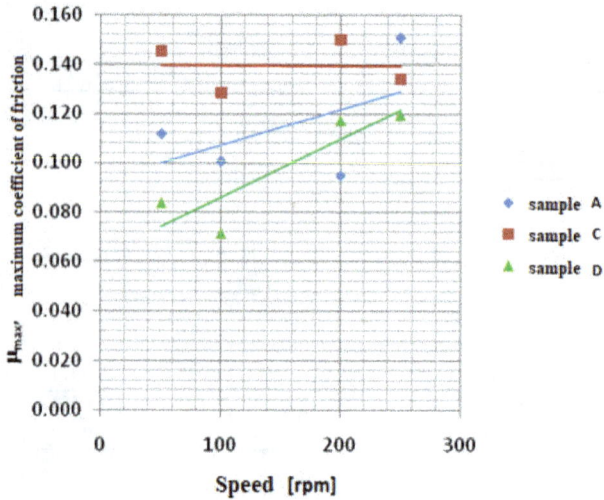

Figure 6.12 Maximum coefficients of friction depending on speed.

The typical variation of the friction moment in time, made in the specialized LabView software before and after the data filtering process is presented in Fig. 6.14.

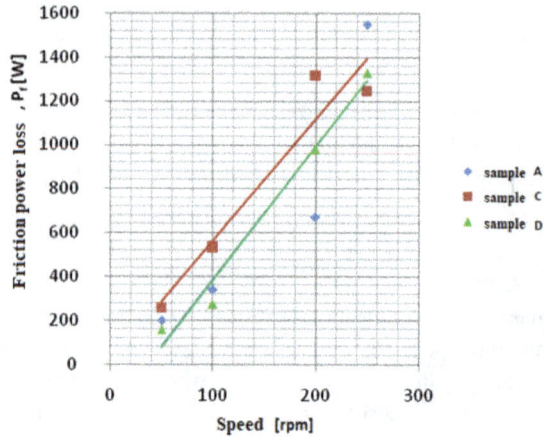

Figure 6.13 Loss of average friction power as a function of speed.

The variation of the signal is due to both the high rotation speed of the wear roller and to the noise of the experimental value-recording equipment.

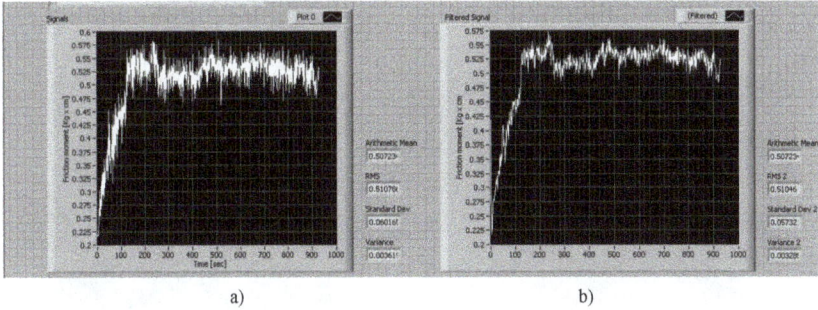

a)   b)

Figure 6.14 Variation of friction torque in LabVIEW: a) before filtration; b) after filtration, for sample D at a speed of 250 [rpm] and a load of 18 N.

The metallic materials were subjected to wear tests described above and their surface was evaluated by electron microscopy and EDS chemical analysis. The aim was to determine the influence of mechanical tests on the microstructure of materials, the chemical composition at the surface and material losses. Fig. 6.15 shows the optical image of the three samples used by semi-dry friction, respectively, D, C and A. Next there are presented the results obtained on the sample C tested with a speed of 200 rotations per minute [14].

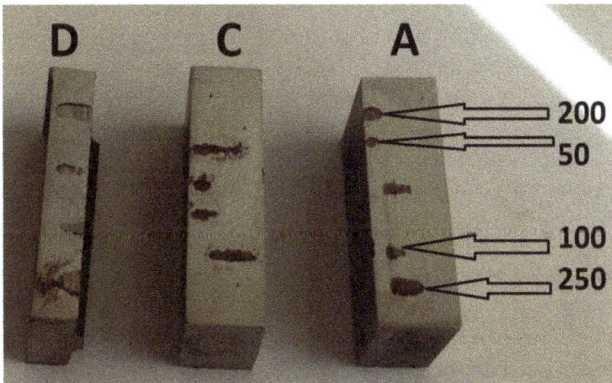

Figure 6.15 Used samples and traces left after the tests with 50, 100, 200 and 250 rpm.

In the wear area, Fig. 6.16 a) and b) was observed the modification of the dendritic phase system under the mechanical action of external forces. In Fig. 6.16 a) is exemplified, by an SEM micrograph, the interface area between a worn surface on the left and a surface only chemically attacked on the right.

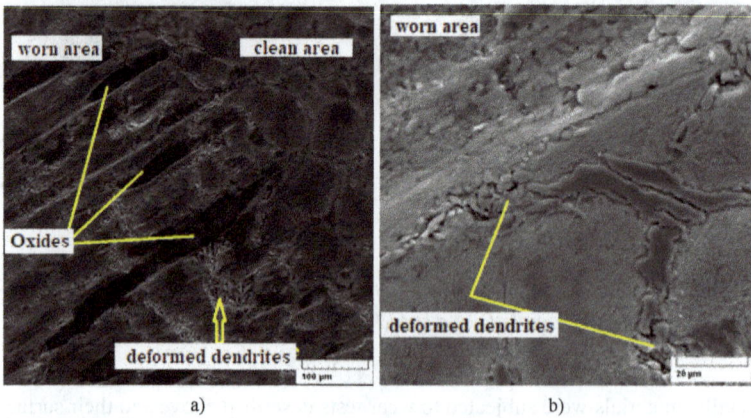

Figure 6.16 SEM images of alloy C (cast iron with ~20% Cr) after mechanical wear: a) 500x; b) 2000x.

It is observed that by applying wear tests, both phases were destroyed on some parts and only the phase of the iron-based matrix in other regions. Cr-based dendrites have better withstood medium and small mechanical stresses, Fig. 6.16 a) and b), due to the different properties of the iron-based phase. Metal carbide-based dendritic compounds behaved differently depending on the force and speed of stress. In this sense, there are areas where the dendrites have completely lost their structural integrity as a result of wear, Fig. 6.16 b) and areas where the dendrites have changed their morphology, mainly shattered into several pieces showing their relatively fragile nature but, clearly, more resistant to mechanical stress than the iron-based matrix.

The presence of carbides in the iron matrix of cast iron changes the distribution of applied external stresses due to the different properties of resistance to stress in the plastic or elastic field. As a direct result, a significant effect can be observed on the wear properties of the material structure, which confirms the results presented above about the coefficient of friction of alloy C. From the microscopies made on the surface, Fig. 6.16 a), an accumulation of material wear, as well as, the appearance of new compounds different from the two basic phases previously characterized.

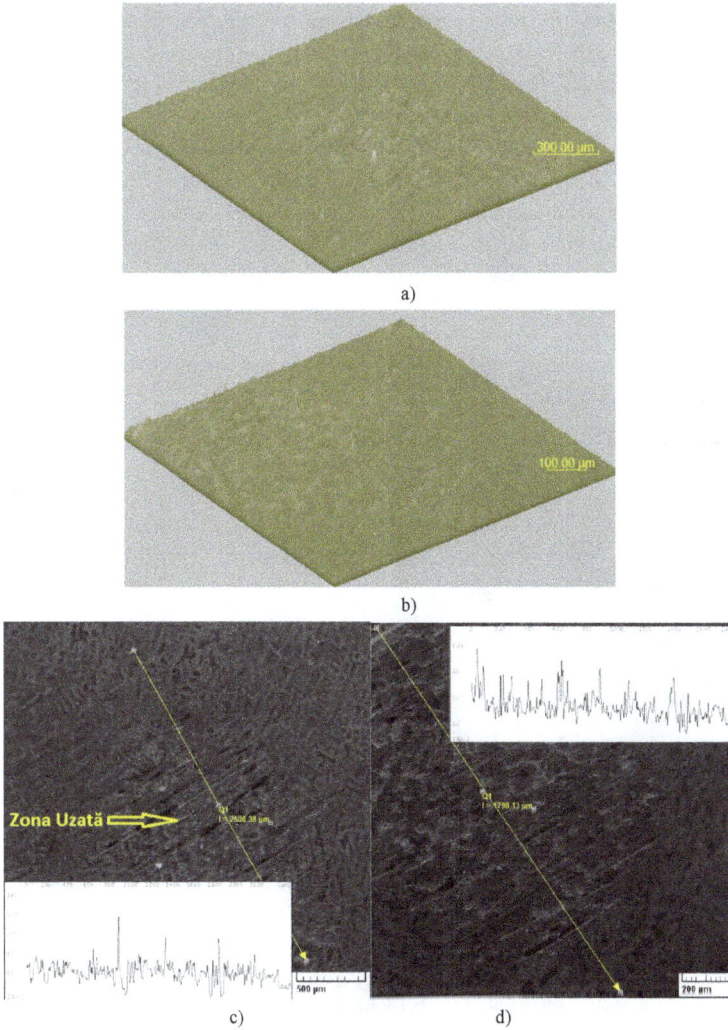

Figure 6.17 Analysis of the surface of alloy D after wear: a) and c) at a speed of 200; b) and d) 250 rpm.

Automotive Brake Disc Materials

Materials Research Forum LLC

Materials Research Foundations **105** (2021)

https://doi.org/10.21741/9781644901458

Fig. 6.17 a) and b) show 3D micrographs made on the surface using the VegaTescan LMHII software for sample D tested with two-cylinder friction speeds of 200 and 250 rpm, respectively. The images show that the wear trace at a speed of 250 rpm is higher but less deep, with a lower loss of material. These results are consistent with the results recorded during the friction test. Fig. 6.17 c) and d) present the surface profiles after the two wear tests by analyzing the variation of light intensity on the surface of the metallic material. Even if the results are dimensionless, there is a variation of wear in the range 60 - 80 for the worn sample at 200 rpm and a variation in the range 50 - 70 for the worn sample at 250 rpm, lower roughness for the case of higher friction speed and small amounts of material lost after the two surfaces slip.

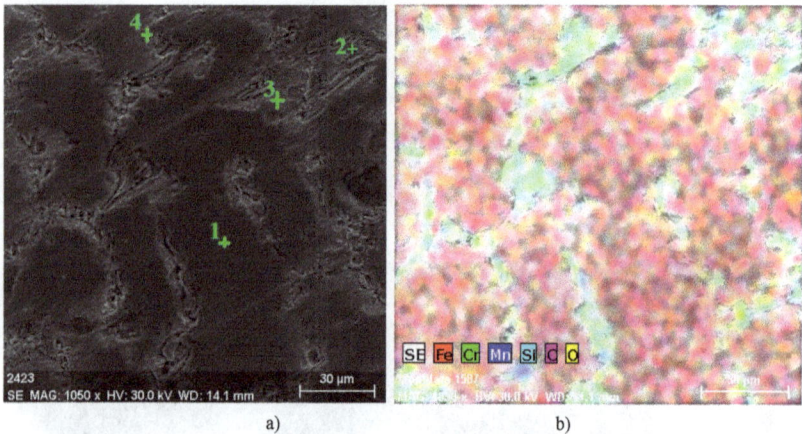

a)   b)

Figure 6.18 Area affected by wear: a) the points selected for the realization of the chemical composition; b) the distribution of the chemical elements Fe, Cr, Mn, Si, C and O on the selected surface.

On the surface affected by wear, Fig. 6.18 a), we selected four points for achieving the chemical composition in areas characteristic of the material, after it has been mechanically worn. It can be appreciated that the surface has suffered, in part, thermal wear with heating of the contact area between the two metallic materials.

In the worn area, Fig. 6.18 b), several oxidized areas are observed, especially in the places affected by wear. The results of the chemical compositions in the 4 selected points are presented in Table 6.4. The manganese element is quantified for the four points selected in

Fig. 6.18 a) at values below 0.1 %wt and was not considered for analysis in these cases. By comparison with the chemical results from the unused sample, we will determine the effects of external mechanical stress on the chemical constituent's characteristic of the C alloy.

Table 6.4 Chemical analysis of the areas selected in points 1-4 of Fig. 6.18 a).

Area	Fe		Cr		C		Si		O	
(Fig.1 a))	wt%	at%	wt%	at%	wt%	at%	wt%	at%	wt%	at%
Point 1	78.4	66.8	16.5	15.1	4.2	16.6	0.9	1.6	-	-
Point 2	27.7	22.8	67	59.2	3.1	11.9	-	-	2.1	6.2
Point 3	74.9	62.8	19.5	17.6	4.1	16	0.7	1.2	0.8	2.4
Point 4	56.58	41.6	31.1	24.6	4.3	14.5	1.2	1.7	6.8	17.6
EDS Error	1.8		0.8		0.7		0.1		0.7	

On the used sample we analyze the chemical composition in four points, three of which are similar to the analyses performed on the unused sample and the fourth on an area with oxides that appeared after the wear tests. In the case of the iron-based matrix, respectively, point 1, no major changes in the chemical composition with respect to the unused material are observed.

In the case of the second surface analyzed, point 2, a loss of the percentage of iron in the dendrites (Cr, Fe) C is observed, based on the loss of material following the mechanical and partial thermal stress of the surface. We notice also an increase in the oxidation of dendrites compared to the initial phase. Simultaneously, some dendritic areas begin more affected by wear and oxidation, point 4 or dark areas on the micrograph in Fig. 6.18 a), where a higher amount of oxygen of up to 6.8 wt% is observed compared to approximately 1 wt% for used samples.

From the microstructural and chemical analysis, it can be seen that the more worn areas are covered by oxide compounds so that we can appreciate that one of the causes of material loss is excessive oxidation of some areas and partial separation of oxides from the base material. The silicon element has not undergone changes in terms of percentage, which is due to the stable and hard compounds it forms, for example different types of carbides.

6.3.2   Analysis of the wear behavior of experimental samples with thin layers

The coated sample was tested on an AMSLER device using a rolling ASTM 52100 steel disc. The data collection system made of a tensometer was used to monitor the friction torque in the tribosystem. A 4-channel Vishay P3 gauge axle was used for data acquisition using the specific software program. The acquired data were processed by the LabVIEW application for virtual signal processing. The mathematical relationships for estimating the friction torque and friction coefficient and the LabVIEW program interface are presented in [14].

A friction test was performed on the AMSLER machine, at a speed of 100 rpm and a constant axial load of approximately 60 N (6kg). The evolution of the friction torque T_f in N x mm and of the coefficient of friction μ is presented in Fig. 6.19. As can be seen, in the first 5 min, the coefficient of friction between the coated layer and the ASTM 52100 steel disc was approximately 0.16 - 0.18, the process of friction and wear being smooth and continuous. After 5 min, the coating was partially removed and a first metal contact with a small surface raised the coefficient of friction to about 0.35, but only for a few seconds. The frictional force of the contact became unstable, but within reasonable limits, until the worn surface expanded and most of the contact became metal on metal.

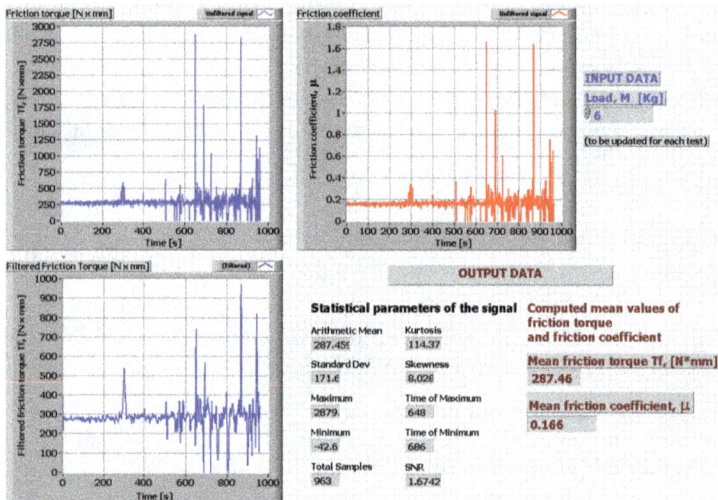

Figure 6.19 The results of the friction test on the AMSLER tribometer.

After 500 seconds from the beginning of the test, the friction became dynamic and this thing can be explained by the intensification of micro-gripping phenomena on the metal contact surface, in the last 5 minutes of testing, the dynamic phenomenon of slipping on the contact area of the tribological contact, manifested by strong vibrations and collisions of the tested samples. Consequently, the variation of the coefficient of friction was enormous. The test was stopped after 15 min due to the simple observation of the wear area on the deposit sample, which reveals the complete removal of the coating from the contact area. Statistical analysis of the data acquisition process (Fig. 6.19) shows that the signal / noise ratio SNR = 1.67, confirming the good quality of the acquired signal. The difficulty of acquisition and standard deviation have high values, confirming the fluctuation of data acquisition at the end of the test, the causes being mentioned above. Comparing the results obtained with those previously reported for EN-GJL-250, a friction coefficient increased around 0.17, can be observed for the whole test. These results recommend Al_2O_3 coatings for applications that require lighter loads.

Future tests must be performed in this direction, with a lower contact pressure and using as roll material, instead of steel roll, ferrule rolls or special materials used in braking systems. Fig. 6.20 shows, through images made by scanning electron microscopy, the wear trace obtained after the test on the Amsler equipment. The trace left is about 4 mm long and 2 mm wide, Fig. 6.20 a). A complete removal of the ceramic layer in the contact area is observed.

The contact made during the experiment was extremely hard because it also engaged material from the substrate during the test, Fig. 6.20 b). The ceramic material was subjected to advanced wear being located between two metallic materials, cast iron as substrate and steel roller for wear. The relatively brittle nature of the layer led to its exfoliation on the contact area, but without further affecting the integrity of the ceramic layer near the wear mark, Fig. 6.20 c).

To highlight the wear area in Fig. 6.21 is presented the distribution of elements Al, O, Fe and C in the contact area after the wear test a) distribution of all elements, b) distribution of aluminum, c) distribution of oxygen and d) distribution of iron.

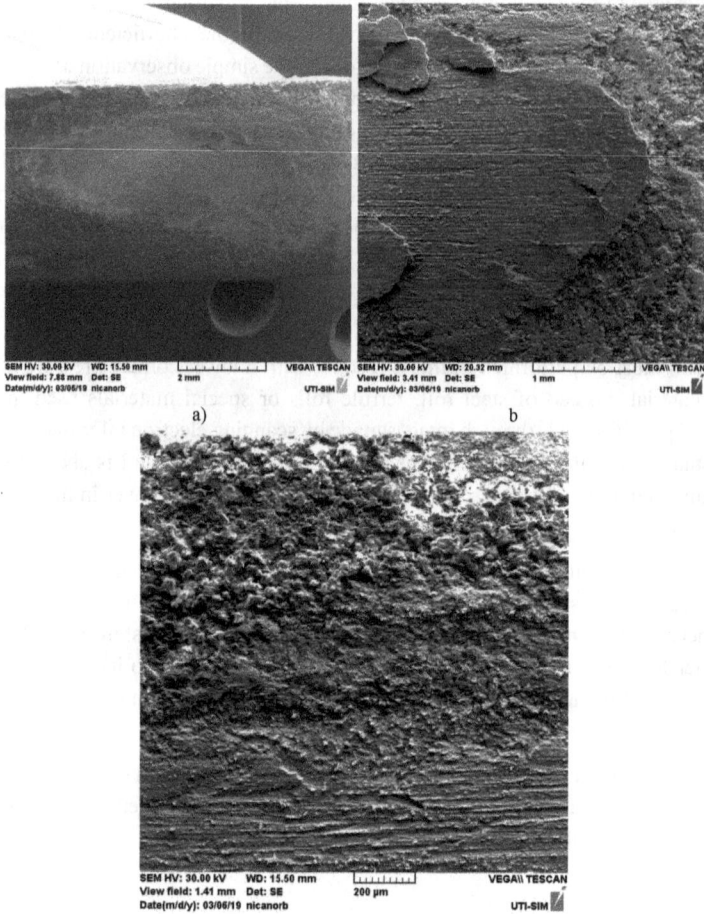

Figure 6.20 SEM images of the area used during the test: a) trace of wear; b) detail of the end of wear; c) the edge of wear.

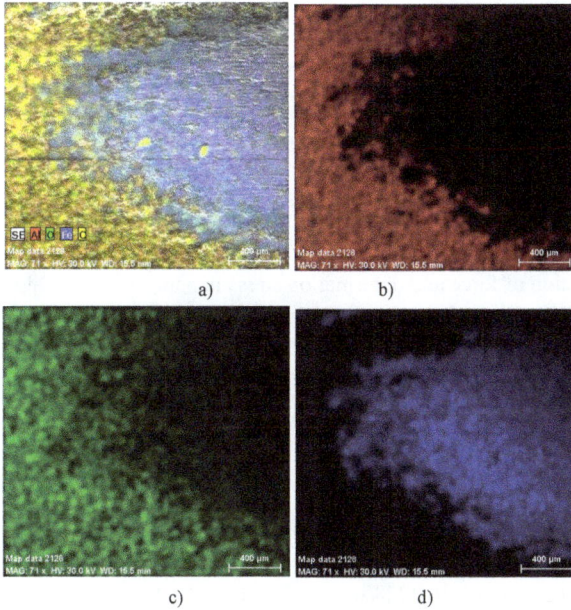

Figure 6.21 Distribution of Al, O, Fe and C elements in the contact area during the wear test: a) distribution of all elements; b) distribution of aluminum; c) oxygen distribution; d) iron distribution.

Observed is the complete removal of the ceramic layer from the contact area. No compacted parts of ceramic material were identified in the contact area. The ceramic layer has exfoliation, micro-cracks and pores only in the areas of the beginning and end of wear, the sides of the contact mark being unaffected by the test. The removal of the ceramic layer was done as a result of strong mechanical shocks that primarily targeted the wear and less the surrounding areas that do not affect the surfaces.

References

[1]   Y. Tan, S. Jiang, D. Yangn, Y. Sheng,. Scratching of Al_2O_3 under pre-stressing, J. Mater. Process. Tech. 211 (2011) 1217–1223. https://doi.org/10.1016/j.jmatprotec.2011.02.005

[2]   C.C. Paleu, C. Munteanu, B. Istrate, S.Bhaumik, P. Vizureanu, M.S. Baltatu, V. Paleu, Microstructural Analysis and Tribological Behavior of AMDRY 1371 (Mo-

NiCrFeBSiC) Atmospheric Plasma Spray Deposited Thin Coatings, Coatings 10 12 (2020) 1186. https://doi.org/10.3390/coatings10121186

[3] J.V.Stebut, F.Lapostolle, M.Bucsa, H.Vallen, Acoustic emission monitoring of single cracking events and associated damage mechanism analysis in indentation and scratch testing. Surface and Coatings Technology. 116–119 (1999) 160–171. https://doi.org/10.1016/S0257-8972(99)00211-X

[4] A.M. Cazac, C. Bejinariu, C. Baciu, S.L. Toma, C.D. Florea, Experimental determination of force and deformation stress in nanostructuring aluminum by multiaxial forging method. Applied Mechanics and Materials. 657(2014) 137-141. https://doi.org/10.4028/www.scientific.net/AMM.657.137

[5] S.Wakayama, K. Ishiwata, Fracture analysis based on quantitative evaluation of microcracking in ceramics using AE source characterization. Journal of Solid Mechanics and Materials Engineering. 3 (2009) 96–105. https://doi.org/10.1299/jmmp.3.96

[6] F. Kaya, Damage detection in fibre reinforced ceramic and metal matrix composites by Acoustic Emission. Key Engineering Materials. 434–435 (2010) 57–60. https://doi.org/10.4028/www.scientific.net/KEM.434-435.57

[7] J. Akbari, Y. Saito, T. Hanaoka, S. Higuchi, S. Enomoto, Effect of grinding parameters on acoustic emission signa ls while grinding ceramics. J. Mater. Process. Tech. 62 (1996) 403–407. https://doi.org/10.1016/S0924-0136(96)02443-0

[8] V. Paleu, G. Gurau, R.I. Comaneci, V. Sampath, C. Gurau,L.G. Bujoreanu, A new application of Fe-28Mn-6Si-5Cr (mass%) shape memory alloy, for self-adjustable axial preloading of ball bearings, Smart Materials and Structures, 27 (2018) 075026. https://doi.org/10.1088/1361-665X/aac4c5

[9] V.P. Singh, A. Sil, R. Jayaganthan, Wear of plasma sprayed conventional and nanostructured Al_2O_3 and Cr_2O_3, based coatings. Chem. Mater. Sci. 65 (2012) 1–12. https://doi.org/10.1007/s12666-011-0070-0

[10] C.D. Florea, C. Bejinariu, C. Savin, B. Istrate, M. Benchea, R. Cimpoesu, , d. Adhesion characterisation of complex ceramics thin layers deposited on metallic substrate, Materials Science Forum. 907 (2017) 126-133. https://doi.org/10.4028/www.scientific.net/MSF.907.126

[11] M. Federici, C. Menapace, A. Moscatelli, S. Gialanella, Effect of roughness on the wear behavior of HVOF coatings dry sliding against a friction material. Wear 368 (2016) 326–334. https://doi.org/10.1016/j.wear.2016.10.013

[12] D.I. Pantelis, P. Psyllaki, N. Alexopoulos,. Tribological behavior of plasma-sprayed Al2O3 coatings under severe wear conditions. Wear. 237 (2000) 197–204. https://doi.org/10.1016/S0043-1648(99)00324-5

[13] H. Czihos, Tribology, Elsevier, Amsterdam, 1978.

[14] C.D.Florea, C. Bejinariu, I. Carcea, V. Paleu, D. Chicet, N. Cimpoeşu, Preliminary results on microstructural, chemical and wear analyze of new cast iron with chromium addition, Key Engineering Materials. 660 (2015) 97-102. https://doi.org/10.4028/www.scientific.net/KEM.660.97

[15] T. Sawaguchi, L.G. Bujoreanu, T. Kikuchi, K. Ogawa, F.X. Yin, Effects of Nb and C in solution and in NbC form on the transformation-related internal friction of Fe-17Mn (mass%) alloys, ISIJ Intern 48 (2008) 99-106. https://doi.org/10.2355/isijinternational.48.99

[16] T. Sawaguchi, L.G. Bujoreanu, T. Kikuchi, K. Ogawa, M. Koyama, M. Murakami, Mechanism of reversible transformation-induced plasticity of Fe-Mn-Si shape memory alloys, Scripta Materialia 59 (2008) 826-829. https://doi.org/10.1016/j.scriptamat.2008.06.030

[17] U. Soyler, B. Ozkal, L. G., Bujoreanu, Sintering Densification and Microstructural Characterization of Mechanical Alloyed Fe-Mn-Si based Powder Metal System, Conference: TMS 2010 Annual Meeting Supplemental Proceedings on Materials Processing and Properties, Seattle, WA,2010 .

[18] G. Gurau, C. Gurau, FMB Fernandes, P. Alexandru, V. Sampath, M. Marin, B.M. Galbinasu, Structural Characteristics of Multilayered Ni-Ti Nanocomposite Fabricated by High Speed High Pressure Torsion (HSHPT), Metals 10 (2020) 1629. https://doi.org/10.3390/met10121629

Chapter 7

Corrosion Resistance of the Materials used for Automotive Brake Disks

R. Cimpoesu[1], C.D. Florea[1], N. Cimpoesu[1], C. Bejinariu[1]*

[1]1Faculty of Materials Science and Engineering, "Gheorghe Asachi" Technical University of Iasi, Romania

costica.bejinariu@tuiasi.ro

Abstract

The materials used in the vehicles braking system are exposed to electrolyte environments such as rainwater, snow and sand, acid rainwater, which can cause significant damage to the brake discs, and reducing their braking capacity. For this reason, the new materials proposed for braking applications must have a higher wear resistance compared to the materials currently used. Cast iron with chromium and cast iron with ceramic deposition were subject of electro-corrosion in rainwater electrolyte solution.

Keywords: Electro-Corrosion, Linear Potentiometry, Tafel, Rainwater, Brake Materials

7.1 Experimental analysis of the corrosion resistance of chromium experimental materials

For the analysis of the corrosion resistance of materials made of standard cast iron (EN-GJL-250) and high cast iron alloyed with chromium, a potentiostat equipment was used. The tests were performed on samples prepared mechanically by grinding in a standard saline solution (0.9% NaCl). The results are more qualitative and only partially quantitative in compliance with the test standards imposed by the G102-89 standard of 2010: Standard practice for calculating the corrosion rate and information obtained from electrochemical measurements. Fig. 7.1 shows the results obtained during the corrosion test, respectively the Tafel linear diagram and the cyclic diagram corresponding to the two analyzed materials: initial cast iron EN-GJL-250 and cast iron with Cr (20%).

From the cyclic diagrams, it is observed that both materials show corrosion in pitting on the entire surface and without large dimensional variations on the analyzed interval [1,2].

The quantitative results recorded on the potentiostat equipment are presented in Table 7.1. Even if the voltage at potential 0, E_0, is close for the two samples, a large difference in polarization resistance is observed, which leads to an increase in the corrosion rate of 2.26 times higher in the case of classical cast iron EN-GJL-250.

Table 7.1 Corrosion resistance analysis parameters recorded during linear and cyclic potentiometry tests.

Sample	E_0 mV	b_a mV	b_c mV	R_p ohm.cm²	J_{cor} mA/cm²	V_{cor} mm/an
EN-GJL-250	-1017.0	660.4	-348.3	323.57	0.1377	3.15
EN-GJL-250+Cr	-1150.2	874.8	-279.3	694.57	0.1312	1.95

Cast iron with a high percentage of chromium (C) has a lower corrosion rate than the initial sample, which is also observed in the Tafel diagrams corresponding to the two samples – Fig. 7.1 a). This increase in corrosion resistance is due to both compounds on chromium base as well as iron matrix passivation in this solution.

Surface analysis of corroded materials was realized by electron microscopy and EDAX chemical analysis. Fig. 7.3 shows the surface conditions of the two materials FC 250 (a and b) and C (20wt% Cr) (c and d). In both cases there is a homogeneous corrosion of the metal surface and the formation of compounds on the surface of the material after the corrosion process. In both cases the structure of the material is highlighted by selective corrosion, especially in the case of cast iron C, of one of the phases. In the case of EN-GJL-250 cast iron, strong corrosion is observed on the entire exposed surface.

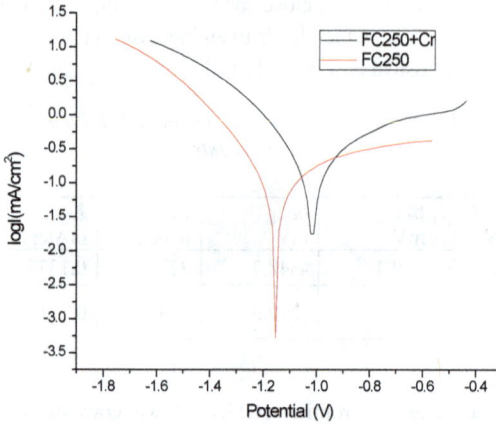

a)

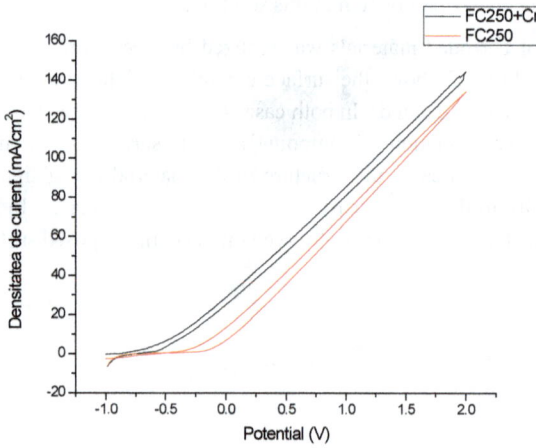

b)

Figure 7.1 Behavior of the two experimental alloys EN-GJL-250 + Cr (C) and EN-GJL-
250 (D) in saline solution: a) Tafel line diagram; b) cyclic diagram.

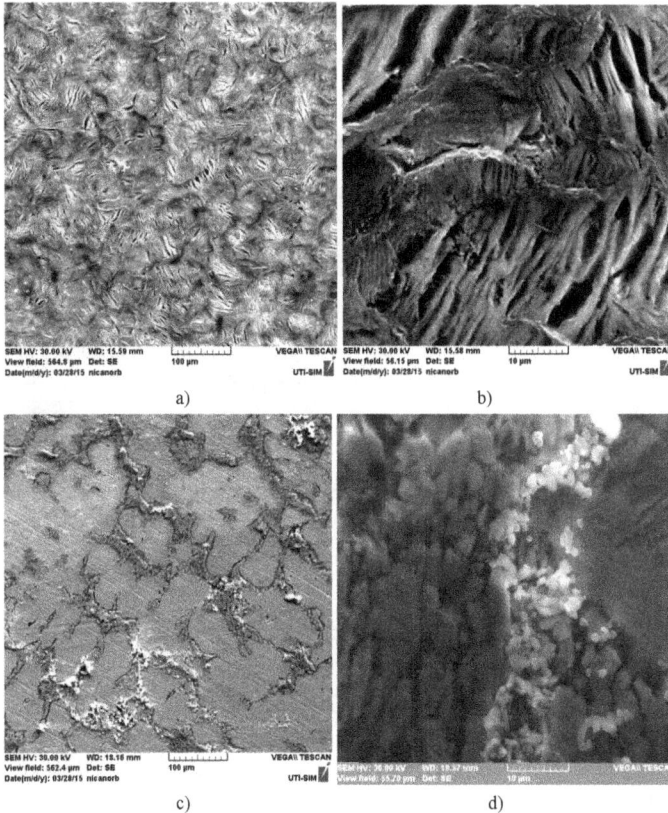

Figure 7.2 SEM images on the surface of corroded materials: a) and b) for standard FC 250 cast iron; c) and d) for cast iron with high Cr content (Cr = 20%, mass percentage).

Corrosion is observed in the case of chromium cast iron, Fig. 7.2 c) and d), especially of the chromium-based dendrites, noting that the basic phase - the iron-based matrix between the Cr dendrites - was relatively little affected. It is observed on the surface of the sample, especially on the dendrites, the formation of metallic compounds, Fig. 7.2 d), with micron or submicron dimensions that have stability on the surface of the metallic material. New compounds are formed mainly from the interface between chromium carbides and the iron-based matrix, Fig. 7.2 d).

In Table 7.2 the energy spectrum characteristic of the identified chemical elements is given qualitatively, and the elements observed on the surface of the initial cast iron EN-GJL-250 after the corrosion test, quantitatively.

Table 7.2 Chemical composition of the cast iron surface EN-GJL-250 (9 mm²) after corrosion test.

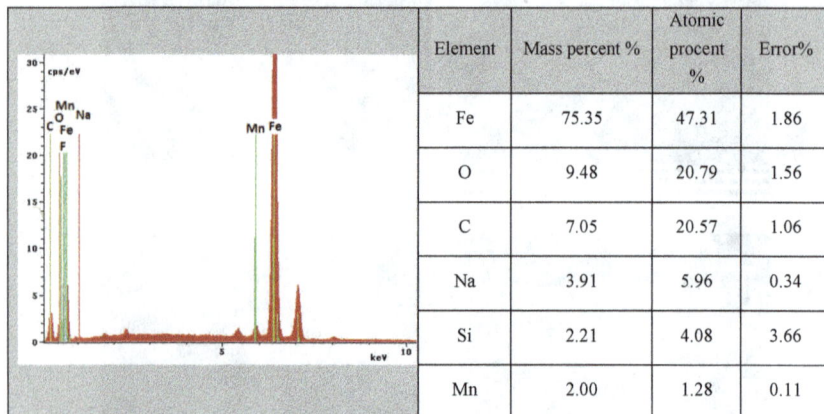

	Element	Mass percent %	Atomic procent %	Error%
	Fe	75.35	47.31	1.86
	O	9.48	20.79	1.56
	C	7.05	20.57	1.06
	Na	3.91	5.96	0.34
	Si	2.21	4.08	3.66
	Mn	2.00	1.28	0.11

Quantitative data are presented in both mass percentages (wt%) and atomic percentages (at%) being mentioned also the error of the EDAX element identification equipment. From the analysis of quantitative data, Table 7.2, there is a considerable loss of iron on the surface of the sample led to an increase in the percentage of other elements in the material: C, Si and Mn, which form more stable and corrosion- resistant compounds. Generally, a significant influence on the corrosion of cast irons has the chlorine ions Cl. The lack of chlorine on the corrosion surface implies the removal of metallic compounds with chlorine in the electrolytic solution.

Fig. 7.3 shows the distribution of some identified elements on the surface of cast iron EN-GJL-250 after corrosion in an area selected with the distribution of all elements in b), Fe in c) and d) Mn, e) Si, f) C, g) Na and h) O. The formation of vermicular graphite and manganese-based compounds is highlighted by corrosion attack. It is observed, in the right part of the distribution in Fig. 7.3 b) and in the distribution in h) the appearance of oxides on the surface of the material confirmed by the high percentage of oxygen identified and presented in Table 7.2.

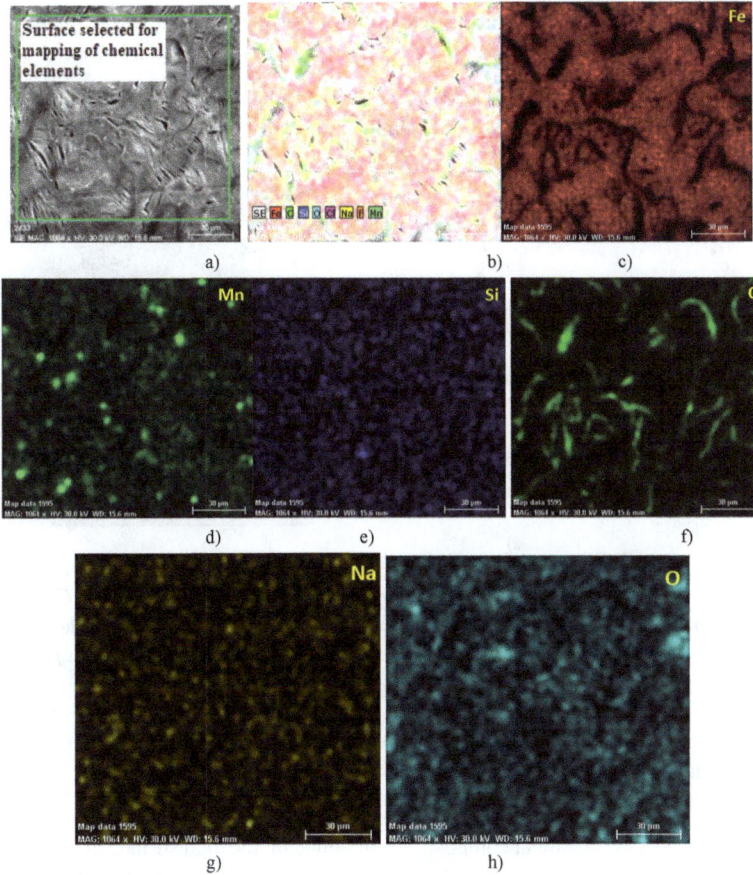

Figure 7.3 Distribution of identified elements on the surface of cast iron EN-GJL-250 after corrosion: a) the area selected for distribution; b) distribution of all elements; c) Fe; d) Mn; e) Si; f) C; g) Na; h) O.

Fig. 7.4 a) shows the energy spectrum characteristic of the elements identified on the surface corroded of chromium metallic material. The presence of the analyzed cast iron characteristic elements (Fe, Cr, C and Si) it is observed on the surface and also other elements that passed from the electrolyte solution on the metallic surface and formed various compounds (Na, O, Cl).

Fig. 7.4 b) shows three areas (1-3) selected for the analysis of the chemical composition in point (90 nm spot) on the surface of the metallic material.

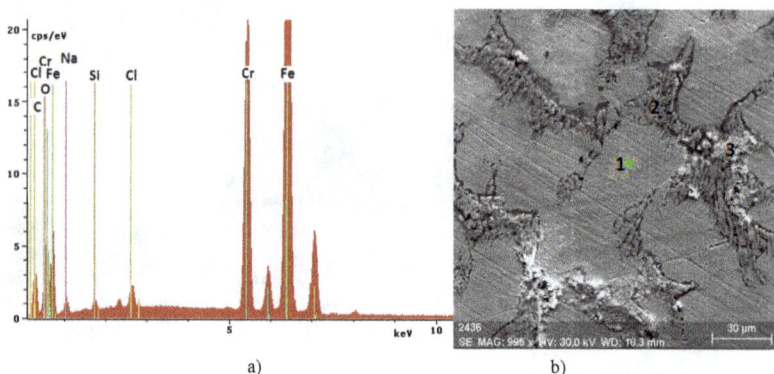

Figure 7.4. Chemical analysis of the surface of the experimental sample C after corrosion: a) the energy spectrum specific to the chemical elements identified on the surface; b) the surface of chemical analysis and the selection of 3 analysis points.

The three areas were selected as follows: point 1 of analysis on the iron-based matrix, point 2 on the dendrite based on (Cr, Fe) C and point 3 on a compound formed after the electro-corrosion test.

Table 7.3 presents the chemical compositions obtained on the total surface in Fig. 7.4 b) as well as in points 1-3 marked on the micrograph. Fig. 7.4 b) highlights the selective attack that occurred in the saline solution on the surface of the iron-based matrix, point 1, being intact from a microstructural viewpoint. Simultaneously, the Cr-based dendrites that were highlighted during the electro-corrosion test are observed. The chemical composition obtained from the general surface of the sample, respectively, area from Table 7.3, is close to the chemical composition obtained on the spark spectrometer and shows traces of new compounds based on O, Na or Cl. From Table 7.3 it is observed that on the iron-based matrix surface, (point 1), there are no traces of oxidation or compounds based on the elements from the solution, which implies a rapid passivation of the material under the conditions of testing in a saline solution, observations confirmed by microstructural analysis.

Table 7.3 Chemical composition of an area of 0.0144 mm² and in the three points selected in Fig. 7.4 b) on a 90 nm spot. All results were achieved in the automatic analysis mode.

	Fe		Cr		Na		O		C		Cl		Si	
	wt %	at%	wt %	at%	wt %	at%	wt %	at%	wt %	at%	wt %	at%	wt %	at%
Area (0.0144 mm²)	61.51	45.89	22.01	17.6	6.6	11.9	5.9	15.4	1.8	6.3	1.2	1.47	0.9	1.3
Point 1 (0.25434 µm²)	78.14	68.6	17.7	16.7	-	-	-	-	3.2	12.9	-	-	1.04	1.8
Point 2 (0.25434 µm²)	33.3	25.9	56.4	47.3	-	-	3.72	10.14	3.47	12.6	3.19	3.92	-	-
Point 3 (0.25434 µm²)	32.9	22.9	50.37	37.8	-	-	9.9	24.2	3.5	11.4	3.3	3.6	-	-
Error EDAX	1.5		0.8		1.2		1.3		0.5		0.1		0.1	

The analyses performed on dendrites, points 2 and 3, show an oxidation of these metallic elements and also the formation of chlorine-based compounds stable on the surface. It is also observed that in addition to a general oxidation of chromium formations there are areas, point 3, in which stable oxides are formed on the surface of the metallic material.

Fig. 7.5 shows the distribution of some elements identified on the surface of cast iron C (Cr = 20 wt%) after corrosion: a) the area selected for distribution, b) the distribution of all elements, c) Fe, d) Cr, e) O, f) Cl, g) C and h) Si.

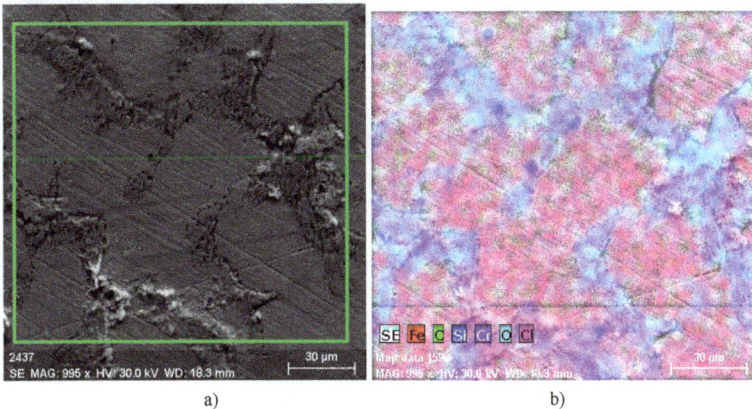

a)   b)

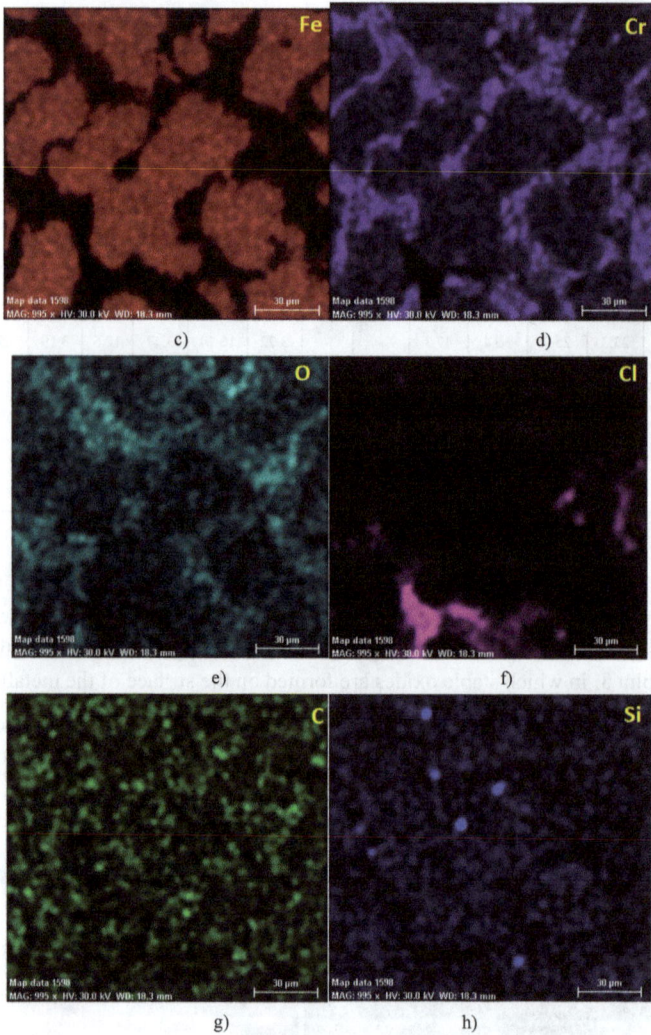

Figure 7.5 Distribution of elements identified on the surface of cast iron C (Cr = 20 wt%) after corrosion: a) the area selected for distribution; b) distribution of all elements; c) Fe; d) Cr; e) O; f) Cl; g) C; h) Si.

In addition to the Fe-based matrix and the Cr-based dendrites highlighted by the distributions in Fig. 7.5 c) and d), the areas with oxides, Fig. 7.5 e) and those with chlorine-based compounds are observed. The intensity of the Cl and Cr signals shows that the thickness of the compound is greater than 5 µm, characteristic that will be analyzed later. We observe that the oxidation is performed mainly on the chromium dendrites.

7.2   Analysis of the surface layers' behavior, deposited on cast iron substrate, at electro-corrosion

Electrochemical tests were performed by linear potentiometry (PGP 201 potentiometer) using a three-electrode cell. Before the experiments, the experimental samples were cleaned for 60 min by ultrasound, in technical alcohol [2,3]. The electrolytic solution used for the experiments was acid rain. The major components of acid rain are the following: sulphuric acid (H_2SO_4), nitric acid (HNO_3) and carbonic acid (H_2CO_3). These chemicals are released into the atmosphere naturally, however, before industrialization, the emergence of factories and dependence on hydrocarbons (coal, petrol, crude oil and others), acid rain was a rare event. In recent decades, acid rain has become an increasingly common occurrence, especially in heavily industrialized areas and crowded cities.

The experimental results show the electro-corrosion resistance of three samples (cast iron substrate EN-GJL-250, EN-GJL-250 + 2 layers of Al2O3 ~ 30 µm ceramic material and EN-GJL-250 layers + 4 layers of Al_2O_3 ~ ceramic material 60 µm) in electrolytic acid rain solution. Fig. 7.6 a) shows the potential-dynamic linear polarization curves of the Al2O3 coating layers with different thicknesses on EN-GJL-250 cast iron compared to the free EN-GJL-250 substrate and in Fig. 7.6 b) cyclic polarization curves.

The linear polarization curves were represented in the potential range: -0.8 - 1 V, using a scanning rate of 1 mV/s. The corrosion rate can be correlated with the intensity of the corrosion current or current density based on Faraday's law. For the experimental cases, corrosion rates of the order of millimeters per year were obtained for the EN-GJL-250 material and micrometers for the coated metallic materials. From Fig. 7.6 a) we see a big difference between the behavior of the cast iron material and that of cast iron with ceramic layers. The difference is not so obvious for cyclic polarization curves, Fig. 7.6 b). Samples coated with ceramic layers show a behavior similar to a non-existent anodic reaction.

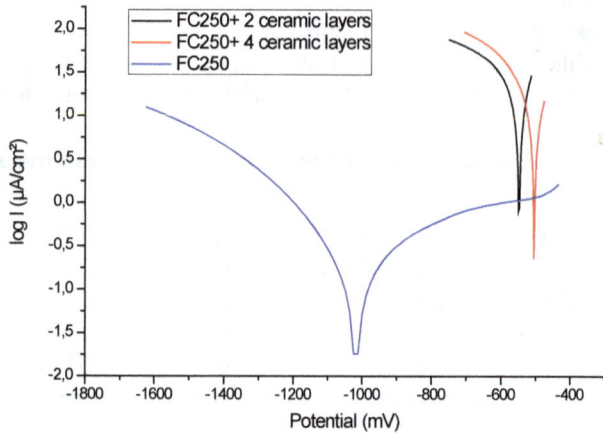

a)

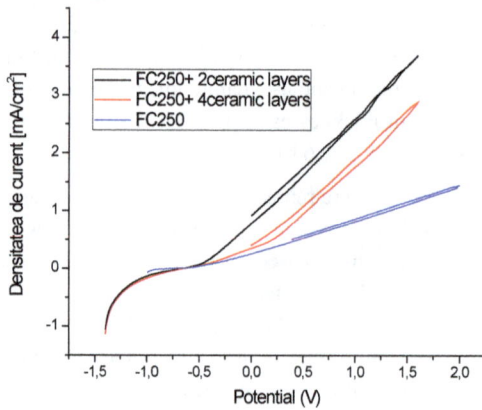

b)

Figure 7.6 Potential-dynamic polarization curves of samples with Al₂O₃ coating layers of different thicknesses on EN-GJL-250 cast iron support compared to the free EN-GJL-250 substrate: a) linear polarization curves; b) cyclic polarization curves.

The cathodic curve of the cyclic curves, Fig. 7.6 b), shows a trajectory similar to the anodic curve - having a reduced hysteresis loop, and the current densities in the passive region are similar to those recorded during direct (anodic) scanning at the same potential [4]. The small difference between the anodic and cathodic line (the lack of a loop) is related to the surface stability and the competition between the diffusion and dissolution in the case of localized corrosion points. Spot corrosion occurs on the basis of a quick diffusion process with a semi-circular dimensional appearance. In the first part of the cathodic process (reversal line), the effects of the dissolution process are reduced, and the time for further diffusion is limited and not sufficient.

The main parameters of the corrosion process (E_0 and j_{corr}) obtained by processing the linear polarization curves are centralized in Table 7.4. The corrosion current thus determined is the corrosion current that occurs at the metal / corrosive medium interface when the metal is introduced into the solution and cannot be measured directly by electrochemical methods. The Open-Circuit Potential (OCP) has large differences between the EN-GJL-250 material and the metallic material with the ceramic layer due to the influence of the inert material layer on the corrosion resistance of the whole assembly. The polarization resistance proved the values of the OCP and are in accordance with the values of the corrosion current [5-8].

The corrosion current of the original material (EN-GJL-250) is 4 to 5 times higher compared to the value recorded for ceramic layer samples. The corrosion rate is 30 to 40 times higher for EN-GJL-250 compared to coated samples.

Table 7.4 Electro-chemical parameters after electro-corrosion tests in the electrolytic solution of acid rain.

Sample	OCP mV	E_0 mV	b_a mV	b_c mV	R_p ohm.cm^2	J_{corr} µA/cm^2	V_{corr} mm/year
EN-GJL-250+2 ceramic layer	-491	548.7	-	-469.5	1450	29.81	0.12
EN-GJL-250+4 ceramic layer	-430	504.6	-	-338.0	1970	25.78	0.10
EN-GJL-250	-716	-1017.0	660.4	-348.3	323.57	137.7	3.64

Scanning electron microscopy (SEM) (VegaTescan LMH II) [9] was used to analyze the morphology of the coatings and the structure of the EN-GJL-250 material before the electrochemical tests and the results are shown in Fig. 7.7a).

In Fig. 7.7 b) and c), the micrographs of the coatings show a dense microstructure with high cohesion and small surface cracks. Moreover, some porous areas are observed in both covered samples. Based on the deposition process, cracks and pores gather and form cracks. The main cause for the appearance of these defects is the short solidification time of the material in the atmosphere and the temperature difference between the deposited layers.

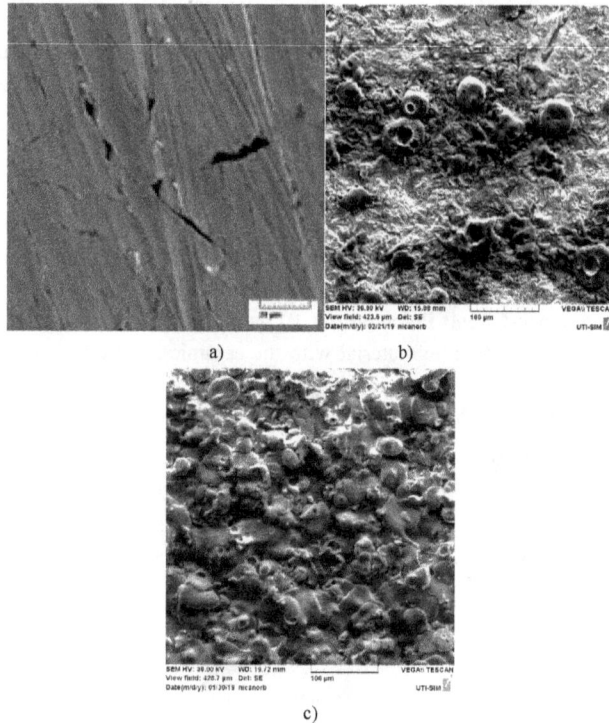

Figure 7.7 SEM images: a) EN-GJL-250 cast iron; b) EN-GJL-250 + 2 layers of ceramic material; c) EN-GJL-250 + 4 layers of ceramic material.

The surface of both deposition cases shows the completely molten surface of the material and the appearance of solidified material formations after the base layers. A relative degree of homogeneity of the coating is crucial for increasing the corrosion resistance of the substrate [10-12].

Fig. 7.7 shows the SEM images of the surface of the experimental materials after the electrochemical tests a) and b) cast iron EN-GJL-250, c) and d) EN-GJL-250 + 2 ceramic layers and e) and f) EN-GJL- 250 + 4 ceramic layers at two different amplifications 200x and 1000x respectively.

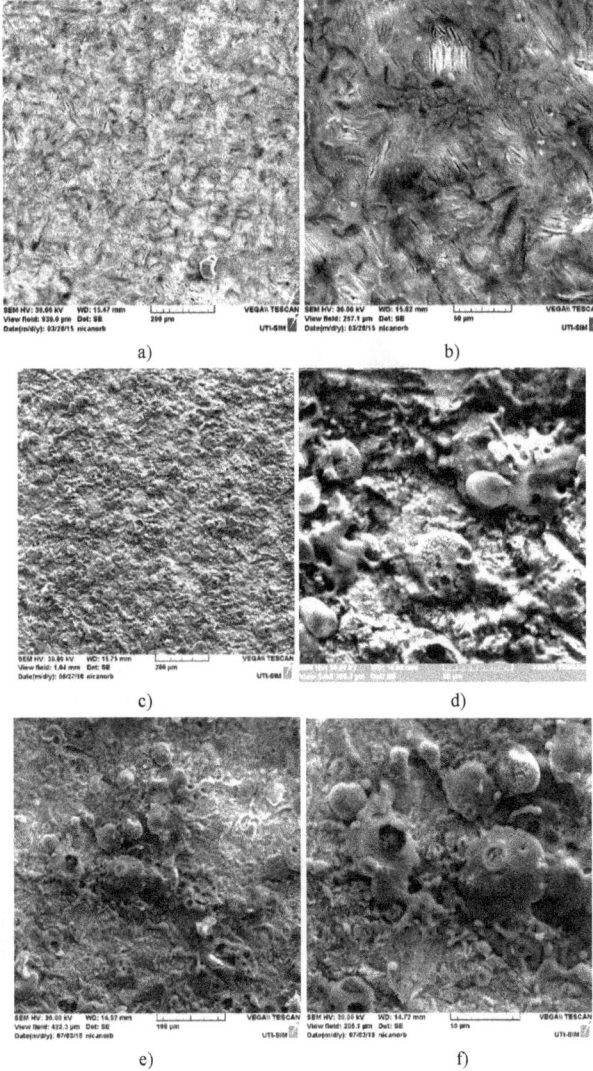

a)

b)

c)

d)

e)

f)

Figure 7.8 Surface SEM images after electrochemical tests: a) and b) cast iron EN-GJL-250; c) and d) EN-GJL-250 + 2 ceramic layers; e) and f) EN-GJL-250 + 4 ceramic layers.

In all cases, Fig. 7.8 confirms the generalized corrosion observed in the cyclic curves, Fig. 7.8 b), without specific areas of corrosion (pitting type spot corrosion). Some coatings in Fig. 7.8 d) and f) show, especially in the areas represented by the particles of ceramic material, a type of pitting corrosion on the outside part of the particles. This behavior does not represent the entire surface, being located only zonally and not being recorded by the potentiostat in the cyclic curves. It can thus be stated that the micro-zonal larger agglomerations on the surface of the layer morphologically, manifest a corrosion behavior in points. If the medium continues to be aggressive (dissolution rate is high enough to overcome diffusion), pitting points on the surface of the ceramic layer can penetrate through the ceramic layer and the electrolyte will come into contact with the metal substrate, which is more susceptible to corrosion [13-17].

In the case of samples covered with ceramic materials, an aggressive surface attack is also observed even if the resistance of the outer oxide layer contributes with a very high resistance, being a material with good chemical inertia. Normally, the inert behavior of ceramic materials that protect the substrate material (such as alumina) keep the metal surface intact. The pores and micro-cracks in the coatings became larger after the electro-corrosion resistance tests because the original micropores were damaged and chemically attacked. The main cause of corrosion is the initial existence of pores and cracks in the coatings. The SEM images shown in Fig. 7.8d) and Fig. 7.8f) suggest that corrosion damage was mainly limited to coating defects (i.e. pores and cracks). It can be seen that some spherical corrosion products have formed around the coating defects.

The results of the EDS analysis, Table 7.5 showed that the corrosion products were mainly composed of Fe and O. It was shown that by electrochemical corrosion ,compounds appeared on the cast iron substrate during the electrochemical experiments.

Table 7.5 Chemical composition of experimental materials after the electro-corrosion resistance test.

Sample	Fe		O		Al		C		Si	
	wt%	at%	wt%	at%	wt%	at%	wt%	at%	wt%	at%
EN-GJL-250	50.56	22.39	42.13	65.13	-	-	3.4	6.9	3.9	3.7
EN-GJL-250+2 layers	32.03	14.7	37.16	49.52	23.93	18.75	4.53	10.85	2.33	1.78

EN-GJL-250+4 layers	33.89	16.77	30.67	45.73	26.93	27.58	4.86	11.12	3.3	3.2
EDS error	0.7		0.95		0.5		0.8		0.1	

The corrosion process occurs mainly through cracks and pores in the ceramic layer that allows the contact of the electrolytic solution with the metal substrate. In all three cases, the materials show an accentuated oxidation of the surface, especially on the EN-GJL-250 material. In the other two experimental cases, it is from the coating and only a percentage participates in the formation of oxides. Generally, the ceramic layer has been penetrated by the electrolyte to the substrate and iron oxides were identified on the surface. Because the ceramic top layer and the metal bonding layer are passive, no much difference in their electrical potentials and no electrical micro-cells are formed between the two materials.

References

[1]   C.D. Florea, I. Carcea, R. Cimpoesu, S.L. Toma, I.G. Sandu, C. Bejinariu, Experimental Analysis of Resistance to Electrocorosion of a High Chromium Cast Iron with Applications in the Vehicle Industry. Rev. Chim.-Bucharest. 68 (2017) 2397-2401. https://doi.org/10.37358/RC.17.10.5893

[2]   N. Aelenei, M. Lungu, D. Mareci, N. Cimpoeşu,. HSLA steel and cast iron corrosion in natural seawater. Environ. Eng. Manag. J., 10 (2011) 1951-1958. https://doi.org/10.30638/eemj.2011.259

[3]   L.C. Trinca, D. Mareci, N. Cimpoesu, M. Calin, T. Stan, Influence of hydrogen peroxide on the corrosion of thermally oxidized ZrTi alloys in phosphate-buffered saline solution. Mater. Corros, 67 (2016) 1088-1095. https://doi.org/10.1002/maco.201508785

[4]   C. Bejinariu, C. Munteanu, C.D. Florea, B. Istrate, N. Cimpoesu, A. Alexandru, A.V. Sandu, Electro-chemical Corrosion of a Cast Iron Protected with a Al2O3 Ceramic Layer. Rev. Chim.-Bucharest. 69 (2018) 3586-3589. https://doi.org/10.37358/RC.18.12.6798

[5]   C.D. Florea, C. Bejinariu, C. Munteanu, B. Istrate, S.L. Toma, A. Alexandru, R. Cimpoesu,. Corrosion Resistance of a Cast-Iron Material Coated With a Ceramic Layer Using Thermal Spray Method. Book Series: IOP Conference Series-Materials Science and Engineering. 374 (2018) UNSP 012028. https://doi.org/10.1088/1757-899X/374/1/012028

[6]   C.D. Florea, C. Munteanu, N. Cimpoesu, I.G. Sandu, C. Baciu, C. Bejinariu, Characterization of Advanced Ceramic Materials Thin Films Deposited on Fe-C Substrate, Rev. Chim.-Bucharest. 68 (2017) 2582-2587. https://doi.org/10.37358/RC.17.11.5933

[7]   C. Bejinariu, D.P. Burduhos-Nergis, N. Cimpoesu, Immersion Behavior of Carbon Steel, Phosphate Carbon Steel and Phosphate and Painted Carbon Steel in Saltwater, Materials 14 (2021) 188. https://doi.org/10.3390/ma14010188

[8]   N. Cimpoesu, F. Sandulache, B. Istrate, R. Cimpoesu, G. Zegan, Electrochemical Behavior of Biodegradable FeMnSi-MgCa Alloy, Metals 8 (2018) 541. https://doi.org/10.3390/met8070541

[9]   M.G. Zaharia, S. Stanciu, R. Cimpoesu, I. Ionita, N. Cimpoesu, Preliminary results on effect of H2S on P265GH commercial material for natural gases and petroleum transportation, Applied surface science 438 (2018) 20-32. https://doi.org/10.1016/j.apsusc.2017.10.093

[10]   J. Izquierdo, G. Bolat, N. Cimpoesu, L.C. Trinca, D. Mareci, R.M. Souto, Electrochemical characterization of pulsed layer deposited hydroxyapatite-zirconia layers on Ti-21Nb-15Ta-6Zr alloy for biomedical application, Applied surface science 385 (2016) 368-378. https://doi.org/10.1016/j.apsusc.2016.05.130

[11]   B. Istrate, J.V. Rau, C. Munteanu, I.V. Antoniac, V. Saceleanu, Properties and in vitro assessment of ZrO2-based coatings obtained by atmospheric plasma jet spraying on biodegradable Mg-Ca and Mg-Ca-Zr alloys, Ceramics international 46 (2020) 15897-15906. https://doi.org/10.1016/j.ceramint.2020.03.138

[12]   I.Gradinaru, I. Stirbu, C.A. Gheorghe, N. Cimpoesu, M. Agop, R. Cimpoesu, C. Popa, Chemical properties of hydroxyapatite deposited through electrophoretic process on different sandblasted samples, Materials Science-Poland 32 (2014) 578-582. https://doi.org/10.2478/s13536-014-0241-x

[13]   B. Istrate, C. Munteanu, R. Cimpoesu, N. Cimpoesu, O.D. Popescu, M. D. Vlad, Microstructural, Electrochemical and In Vitro Analysis of Mg-0.5Ca-xGd Biodegradable Alloys, Applied Sciences-Basel 11 (2021) 981. https://doi.org/10.3390/app11030981

[14]   D.P. Burduhos-Nergis, P. Vizureanu, A.V. Sandu, C. Bejinariu, Phosphate Surface Treatment for Improving the Corrosion Resistance of the C45 Carbon Steel Used in Carabiners Manufacturing, Materials, 13 (2020) 3410. https://doi.org/10.3390/ma13153410

[15]   C. Nejneru, M.C. Perju, D.D.B. Nergis, A.V. Sandu, C. Bejinariu, Galvanic Corrosion Behaviour of Phosphate Nodular Cast Iron in Different Types of Residual Waters and Couplings, Revista De Chimie 70 (2019) 3597-3602. https://doi.org/10.37358/RC.19.10.7604

[16]   A.V. Sandu, M.S. Baltatu, M. Nabialek, A. Savin, P. Vizureanu, Characterization and Mechanical Proprieties of New TiMo Alloys Used for Medical Applications, Materials, 12 (2019) 2973. https://doi.org/10.3390/ma12182973

[17]   D.P. Burduhos-Nergis, P. Vizureanu, A.V. Sandu, C. Bejinariu, Evaluation of the Corrosion Resistance of Phosphate Coatings Deposited on the Surface of the Carbon Steel Used for Carabiners Manufacturing, Applied Sciences-Basel 10 (2020) 2753. https://doi.org/10.3390/app10082753

About the Authors

Costel Dorel FLOREA

PhD Eng. with 10 ISI papers and more than 50 citations. Co-organizer of Safety in Industrial field Congres. The main contributions focused on improving the properties of the usual materials used for brake discs by alloying or by deposition of thin ceramic layers on the contact surface. He characterized the classic materials used to make brake discs and improving the mechanical and chemical characteristics of existing materials.

Nicanor CIMPOESU

Professor Ph.D. Eng.
Materials Science and Engineering Faculty,
"Gheorghe Asachi" Technical University of Iasi
nicanor.cimpoesu@tuiasi.ro

Professor and researcher at "Gheorghe Asachi" Technical University of Iasi, with more than 15 years of experience. Prof. Habilitated since 2019, with 5 undergoing PhD students.Coordinator of Energy Spectroscopy and Image Microscopy (ESIM) laboratory from 2007 (esim.sim.tuiasi.ro). Field of experience is Materials Engineering with 12 published books, 120 ISI published articles with more than 750 citations – H index of 15/WoS and 16/Scopus. Worked on more than 25 research grants, on 2 being director and another 1 institution responsible. He has 3 patents and many awards received for them. He is a member of EMS, SMER and ATTR academic societies and also reviewers for many scientific journals, conferences and projects.

Costica BEJINARIU

Professor Ph.D. Eng.
Vicedean of Faculty of Materials Science and Engineering,
"Gheorghe Asachi" Technical University of Iasi
costica.bejinariu@tuiasi.ro

Professor and researcher at "Gheorghe Asachi" Technical University of Iasi, with more than 30 years of experience. PhD Coordinator since 2009, with 5 granted PhD students and 8 undergoing PhD student. Field of experience is Materials Engineering with 20 published books over 200 published articles with more than 600 citations – H index of 15/WoS and 18/Scopus. Worked on more than 45 research grants, on 4 being director and another 3 institution responsible. He has 12 patents and many awards received for them. He is a member of various academic societies and also reviewers for many scientific journals and conferences.

Ramona CIMPOESU

Lecturer Ph.D. Eng. Chem.
Materials Science and Engineering Faculty,
"Gheorghe Asachi" Technical University of Iasi
ramona.cimpoesu@tuiasi.ro

Lecturer at Materials Science Departament and responsible of Corrosion resistance laboratory, with more than 10 years of experience. She is the author of more than 50 ISI articles, Hirsh index 8 and more than 200 citations. The significant scientific contribution to most articles is the analysis of metallic materials in terms of corrosion resistance. She analyzed the corrosion resistance by electrochemical methods: linear and cyclic potentiometry and by electrochemical impedance spectrometry. Co-author at 5 books in technical field. Member of AGIR society. She participated as a specialist member in the research team in many scientific research contracts. Reviewer for many international Journals.

www.ingramcontent.com/pod-product-compliance
Lightning Source LLC
Chambersburg PA
CBHW071703210326
41597CB00017B/2313